The Open University

M381 Number Theory and
Mathematical Logic

GW00598032

Number Theory **Unit 4**

Fermat's and Wilson's Theorems

Prepared for the Course Team by Alan Best

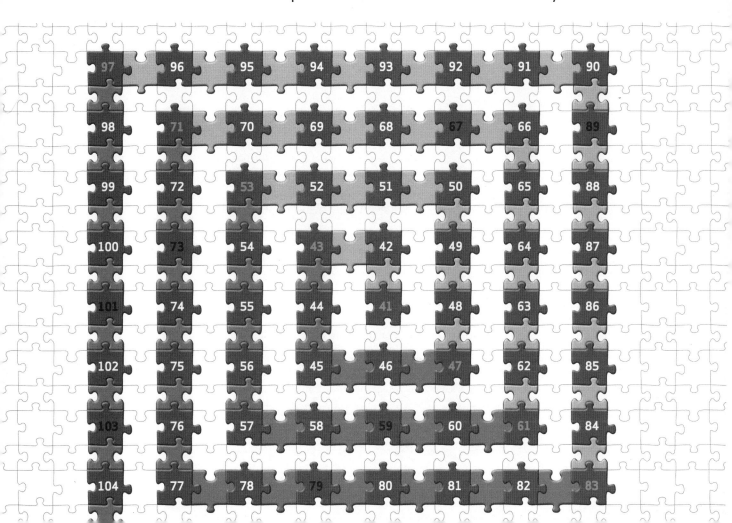

The M381 Number Theory Course Team

The Number Theory half of the course was produced by the following team:

Alan Best	*Author*
Andrew Brown	*Course Team Chair* and *Academic Editor*
Roberta Cheriyan	*Course Manager*
Bob Coates	*Critical Reader*
Dick Crabbe	*Publishing Editor*
Janis Gilbert	*Graphic Artist*
Derek Goldrei	*Critical Reader*
Caroline Husher	*Graphic Designer*
John Taylor	*Graphic Artist*

with valuable assistance from:

CMPU	*Mathematics and Computing, Course Materials Production Unit*
John Bayliss	*Reader*
Elizabeth Best	*Reader*
Jeremy Gray	*History Reader*
Alison Neil	*Reader*

The external assessor was:

Alex Wilkie	*Reader in Mathematical Logic, University of Oxford*

This publication forms part of an Open University course. Details of this and other Open University courses can be obtained from the Student Registration and Enquiry Service, The Open University, PO Box 197, Milton Keynes, MK7 6BJ, United Kingdom: tel. +44 (0)870 300 6090, e-mail general-enquiries@open.ac.uk

Alternatively, you may visit the Open University website at http://www.open.ac.uk where you can learn more about the wide range of courses and packs offered at all levels by The Open University.

To purchase a selection of Open University course materials, visit http://www.ouw.co.uk, or contact Open University Worldwide, Michael Young Building, Walton Hall, Milton Keynes, MK7 6AA, United Kingdom, for a brochure: tel. +44 (0)1908 858793, fax +44 (0)1908 858787, e-mail ouw-customer-services@open.ac.uk

The Open University, Walton Hall, Milton Keynes, MK7 6AA.

First published 1996. Reprinted 2001 and 2003. New edition 2007 with corrections.

Copyright © 1996 The Open University

Edited, designed and typeset by The Open University, using the Open University TeX System.

Printed and bound in the United Kingdom by The Charlesworth Group, Wakefield.

ISBN 978 0 7492 2270 3

2.1

CONTENTS

INTRODUCTION

One way to investigate problems in number theory is to look at special cases of the problem and to draw general conclusions from the empirical evidence accumulated. We have already seen instances, and we shall see many more, of dangers in this approach when seemingly clear patterns of behaviour can go unexpectedly wrong. Consider the numbers $2^n - 2$, for positive integers n. Investigation of the prime decompositions of these numbers would yield that $2^2 - 2$ is divisible by 2, $2^3 - 2$ is divisible by 3, $2^5 - 2$ is divisible by 5 and $2^7 - 2$ is divisible by 7. On the other hand taking composite exponents, $2^4 - 2$ is not divisible by 4 and $2^6 - 2$ is not divisible by 6. It is beginning to look as if $2^n - 2$ is divisible by n when n is prime but not when it is composite. Trying more and more values of n, we find that $2^{17} - 2$ is divisible by 17, $2^{18} - 2$ is not divisible by 18, $2^{19} - 2$ is divisible by 19, and this no doubt fortifies our faith in the truth of this result. But no matter how much numerical evidence is gathered, we cannot be sure that the pattern really does persist until we have a general proof.

For many hundreds of years mathematicians believed the result that $2^p - 2$ is divisible by p if, and only if, p is prime. They had good evidence to support their conjecture, but not a proof. It was 1819 before a counterexample was discovered. $341 = 11 \times 31$ is not prime but $2^{341} - 2$ is divisible by 341.

In the 17th century Fermat showed interest in these numbers and he extended his researches beyond just looking at powers of 2. His interest was captured by numbers $a^p - a$ for any integer a, and he succeeded in proving that if p is prime then $a^p - a$ is divisible by p. He offered no conclusions when the exponent is composite.

In his book *Men of Mathematics* (Pelican), E.T. Bell writes

> It is difficult if not impossible to state why some theorems in arithmetic are considered 'important' while others, equally difficult to prove, are dubbed trivial. One criterion, although not necessarily conclusive, is that the theorem shall be of use in other fields of mathematics. Another is that it shall suggest researches in arithmetic or in mathematics generally, and a third that it shall be in some respect universal. Fermat's Theorem satisfies all of these somewhat arbitrary demands: it is of indispensable use in many departments of mathematics, including the theory of groups, which in turn is at the root of the theory of algebraic equations; it has suggested many investigations, of which the entire subject of primitive roots may be recalled to mathematical readers as an important instance; and finally it is universal in that it states a property of *all* prime numbers — such statements are extremely difficult to find and very few are known.

When, in the first section of this unit, we meet Fermat's Theorem, which tends to be known as Fermat's Little Theorem to distinguish it from Fermat's Last Theorem which we shall meet in *Unit 8*, it may not appear to be anything remarkable. Yet as the unit unfolds we shall show many applications of the theorem, and in particular we shall capitalize on its immense potential in assisting in calculations involving large numbers. Deservedly it is considered to be a cornerstone in the development of number theory.

The role of Fermat's Theorem in solving certain problems will lead us into the subject of primitive roots (referred to in the extract from Bell above). We shall pursue these ideas more comprehensively in the next unit, but here we shall use them to explain the periodic behaviour of the decimal expansions of fractions.

This unit introduces a second important result, Wilson's Theorem, which is similar in nature to Fermat's Little Theorem although it is not connected in any way. This theorem asserts that if p is prime then $(p-1)! \equiv -1 \pmod{p}$. Wilson's Theorem shares with Fermat's Little Theorem the feature that both are concerned with properties of *all* prime numbers. What is more they have another rare feature in common: for both there is a relatively simple general proof available!

The statements of both Fermat's and Wilson's Theorems are implications, that is, they are of the form *if A holds then B holds*. For such statements it is natural to ask about the converse: *if B holds then A holds*. The converse of Wilson's Theorem, which states that for $p > 1$, if $(p-1)! \equiv -1 \pmod{p}$ then p is prime, will be seen to be true. On the other hand the converse of Fermat's Little Theorem will be seen to be false. It is not true that if, for all integers a, we have that $a^p - a$ is divisible by p, then p must be prime.

1 FERMAT'S LITTLE THEOREM

1.1 Statement and proof of Fermat's Little Theorem

Figure 1.1 below shows successive powers of integers representing the non-zero residue classes modulo the respective primes 3, 5 and 7.

mod 3		
a	1	2
a^2	1	1

mod 5				
a	1	2	3	4
a^2	1	4	4	1
a^3	1	3	2	4
a^4	1	1	1	1

mod 7						
a	1	2	3	4	5	6
a^2	1	4	2	2	4	1
a^3	1	1	6	1	6	6
a^4	1	2	4	4	2	1
a^5	1	4	5	2	3	6
a^6	1	1	1	1	1	1

Figure 1.1   Powers modulo a prime

There are some interesting patterns beginning to emerge from these tables. Notice how some power of each element is congruent to 1, and thereafter the powers 'cycle'. For instance, if we continue to determine more and more powers of 4 modulo 7 then we get the sequence

$$4, 2, 1, 4, 2, 1, 4, 2, 1, 4, 2, 1, \ldots,$$

with the cycle 4, 2, 1 of length 3.

The powers of 5 modulo 7 give the sequence

$$5, 4, 6, 2, 3, 1, 5, 4, 6, 2, 3, 1, \ldots,$$

with a cycle 5, 4, 6, 2, 3, 1 of length 6.

In particular, notice how the last listed row of each table consists entirely of 1's. That is, for these three prime moduli, $a^{p-1} \equiv 1 \pmod{p}$ for each $a \not\equiv 0 \pmod{p}$. The result illustrated here holds generally for any prime p, and was first formulated by Fermat. This result may look slightly different from the claim in the introduction that, for all integers a, $a^p - a$ is divisible by p, but we shall see shortly that they are indeed the same.

> **Theorem 1.1 Fermat's Little Theorem**
>
> If p is a prime and a is any integer with $\gcd(a, p) = 1$, then
>
> $$a^{p-1} \equiv 1 \pmod{p}.$$

The condition $\gcd(a, p) = 1$ is equivalent to each of $a \not\equiv 0 \pmod{p}$ and 'p does not divide a'. Depending on the application we shall use the most appropriate form.

Proof of Theorem 1.1

Consider the set of $p - 1$ integers

$$\{a, 2a, 3a, \ldots, (p-1)a\}.$$

None of these numbers is congruent modulo p to 0, for if $ra \equiv 0 \pmod{p}$, Euclid's Lemma gives $r \equiv 0 \pmod{p}$ or $a \equiv 0 \pmod{p}$, neither of which is the case here. What is more, no two of these numbers are congruent to each other modulo p, for if we had

$$ra \equiv sa \pmod{p}, \text{ where } 1 \leq r \leq p - 1 \text{ and } 1 \leq s \leq p - 1,$$

then cancellation of the a, which we can do since $\gcd(a, p) = 1$, gives $r \equiv s \pmod{p}$ and so $r = s$. It follows that these $p - 1$ numbers are congruent, in some order, to the numbers in the set

$$\{1, 2, 3, \ldots, p - 1\}$$

consisting of the least positive residues of p, excluding zero.

So, multiplying all the numbers in each set together,

$$a \times 2a \times 3a \times \cdots \times (p-1)a \equiv 1 \times 2 \times 3 \times \cdots \times (p-1) \pmod{p},$$

which gives

$$a^{p-1} \times (p-1)! \equiv (p-1)! \pmod{p}.$$

Since p does not divide $(p-1)!$ the latter can be cancelled throughout the congruence leaving $a^{p-1} \equiv 1 \pmod{p}$, as required. ∎

Pierre de Fermat (1601–1665)

The 17th century was a fruitful period for mathematics. Descartes was the first to apply algebra to geometry thereby giving birth to the subject of analytical geometry, Pascal created the mathematical theory of probability and Newton and Leibnitz were developing the calculus. However, it is often argued that the greatest mathematician of this period was Fermat.

Fermat was born near Toulouse, France, the son of a leather merchant. He lived a quiet, but eventful life in and around his birthplace. For his entire working life of 34 years he was a lawyer and magistrate in the local parliament of Toulouse. Fermat had no mathematical training and, indeed, showed no apparent interest in the subject until he was beyond 30 years of age. He took to mathematics as a recreation, for the sheer love of it. An amateur he may have been, but his contributions to many areas of mathematics make him indisputably one of the all time greats.

Fermat is regarded as a co-inventor of analytical geometry, working independently of Descartes on applications of algebra to geometry. One offshoot of his work was that he discovered a method for finding maxima and minima of functions well before Newton and Leibnitz came on the scene. Indeed a letter from Newton acknowledges that he got hints for his differential calculus from Fermat's method for drawing tangents.

But Fermat's first love, and his greatest work, was in number theory, or 'higher arithmetic' as it was then known. He was at his best in tackling problems involving primes and solving problems involving large numbers requiring vast amounts of computation. However, it was his use of new principles and methods which most influenced following generations and which could be said to have been the birth of modern number theory.

Being an amateur, Fermat did not seek a reputation for his mathematical achievements. He refused to have any of his work published, preferring instead to correspond with a few contemporaries, most of whom were fellow amateurs. Our knowledge of Fermat's works relies on these letters and on the preservation of notes collected by his family after his death. The margins of his copy of a book on arithmetic, *Bachet's Diophantus*, contained many of Fermat's results in number theory. The limited amount of space in the margin led to Fermat's habit of quoting results, but omitting key steps in the derivation of the result. Many is the time that mathematicians have wished that the margins of his books had been wider!

Fermat explained the result of his Little Theorem in a letter to an official at the French mint, Frenicle. The letter included the comment, 'I would send you the demonstration if I did not fear its being too long'. Although it is generally accepted that Fermat did possess such a demonstration (or proof), nearly 100 years passed before a proof of Fermat's Little Theorem was at last published, by Euler.

It is certainly not true that $a^{p-1} \equiv 1 \pmod{p}$ when $a \equiv 0 \pmod{p}$, for $0^{p-1} \equiv 0 \pmod{p}$. However Fermat's Little Theorem, (which we shall refer to in the abbreviated form FLT), can be expressed in an alternative way which includes the case $a \equiv 0 \pmod{p}$, as follows.

Fermat's Little Theorem, an alternative formulation

If p is a prime and a is any integer,

$$a^p \equiv a \pmod{p}.$$

Problem 1.1 _____

Prove that the two formulations of FLT are equivalent. That is, prove that for prime p:

(a) if $a^p \equiv a \pmod{p}$ for all integers a, then $a^{p-1} \equiv 1 \pmod{p}$ for all $a \not\equiv 0 \pmod{p}$;

(b) if $a^{p-1} \equiv 1 \pmod{p}$ for all $a \not\equiv 0 \pmod{p}$, then $a^p \equiv a \pmod{p}$ for all integers a.

FLT has applications in many areas, as we shall see. But it really comes into its own in assisting with complex calculations, as in the following examples.

Example 1.1

What is the remainder when 7^{40} is divided by 17?

FLT tells us that $7^{16} \equiv 1 \pmod{17}$. Therefore, breaking the power 40 down in an appropriate way,

$$7^{40} = (7^{16})^2 \times 7^8 \equiv 1^2 \times 7^8 \equiv 7^8 \pmod{17}.$$

Now $7^2 = 49 \equiv -2 \pmod{17}$, and so

$$7^{40} \equiv 7^8 \equiv (7^2)^4 \equiv (-2)^4 \equiv 16 \pmod{17},$$

and the remainder on dividing 7^{40} by 17 is 16. ◆

Example 1.2

Show that, for any odd prime p,

$$1^p + 2^p + 3^p + \cdots + (p-1)^p \text{ is divisible by } p.$$

By FLT, $a^p \equiv a \pmod{p}$ for all integers a. Therefore

$$1^p + 2^p + 3^p + \cdots + (p-1)^p \equiv 1 + 2 + 3 + \cdots + (p-1) \pmod{p}$$

$$\equiv \frac{p(p-1)}{2} \pmod{p}, \quad \text{sum of an arithmetic progression}$$

$$\equiv 0 \pmod{p}, \quad \text{since } \frac{p-1}{2} \text{ is an integer when } p \text{ is odd.}$$

So p divides $1^p + 2^p + 3^p + \cdots + (p-1)^p$ for all odd primes p. ◆

Problem 1.2 _____

Find the remainder when

(a) 5^{20} is divided by 7;

(b) 37^{37} is divided by 17.

Problem 1.3 _____

Let p be a prime and $\gcd(a, p) = 1$. Use FLT to verify that
$x \equiv a^{p-2}b \pmod{p}$ is a solution of the linear congruence $ax \equiv b \pmod{p}$.
Hence solve $5x \equiv 18 \pmod{19}$.

1.2 Pseudoprimes

The converse of the alternative formulation of FLT fails to hold. That is, if
$a^n \equiv a \pmod{n}$ for every integer a, it does not follow that n is prime. But
counterexamples to this converse are not so easy to find. Concentration on
the case $a = 2$ has historical interest. For a very long time it was thought n
had to be prime for $2^n - 2$ to be divisible by n, and this was used as a test
for the primality of n. Let us first put an end to that conjecture.

Example 1.3

Show that $2^{341} \equiv 2 \pmod{341}$.

We shall show that $2^{340} \equiv 1 \pmod{341}$, from which the result follows on
multiplying through by 2.

Notice that $341 = 11 \times 31$, so our line of attack will be to determine first the
values of $2^{340} \pmod{11}$ and $2^{340} \pmod{31}$. As 11 and 31 are primes we can
call on FLT; this tells us that $2^{10} \equiv 1 \pmod{11}$ and that $2^{30} \equiv 1 \pmod{31}$.
Working modulo 11:

$$2^{340} \equiv (2^{10})^{34} \equiv 1^{34} \equiv 1 \pmod{11}.$$

341 is not a counterexample to the
converse of FLT since it is not true
that $a^{341} \equiv a \pmod{341}$ for *all*
integers a. For example,
$3^{341} \equiv 168 \pmod{341}$.

Working modulo 31:

$$2^{340} \equiv (2^{30})^{11} \times (2^{10}) \equiv 1^{11} \times 2^5 \times 2^5 \equiv 1 \times 1 \times 1 \equiv 1 \pmod{31}.$$

As 11 and 31 are relatively prime we can appeal to Theorem 1.3 of *Unit 3* to
conclude, from the above two congruences, that

$$2^{340} \equiv 1 \pmod{341}.$$ ◆

Composite integers n with the property that n divides $2^n - 2$ are called
pseudoprimes. They do not have to be odd numbers; the two smallest even
pseudoprimes are $161\,038$ and $215\,326$. There are infinitely many
pseudoprimes, although they appear to be much sparser than the primes
themselves. For example, a recent calculation has shown that of the numbers

Pseudoprimes are sometimes called
Poulet numbers, after the French
mathematician who, in 1926,
computed them all up to 5×10^7.

up to 2×10^{10} there are $882\,206\,716$ primes but only $19\,865$ pseudoprimes. The smallest pseudoprime, discovered in 1819, is 341. The next, discovered in 1912, furnishes us with the first counterexample to the converse of FLT.

Example 1.4

Show that $a^{561} \equiv a \pmod{561}$ for every integer a.

First, we factorize 561 as $561 = 3 \times 11 \times 17$. We shall determine the value of a^{561} modulo each of 3, 11 and 17, using FLT. For any integer a,

$$a^3 \equiv a \pmod{3},\ a^{11} \equiv a \pmod{11} \text{ and } a^{17} \equiv a \pmod{17}.$$

Working modulo 3:

$$a^{561} = (a^3)^{187} \equiv a^{187} = (a^3)^{62} \times a \equiv a^{62} \times a \equiv a^{63} \equiv (a^3)^{21} \equiv a^{21}$$
$$\equiv (a^3)^7 \equiv a^7 \equiv (a^3)^2 \times a \equiv a^2 \times a \equiv a^3 \equiv a \pmod{3}.$$

Working modulo 11:

$$a^{561} = (a^{11})^{51} \equiv a^{51} = (a^{11})^4 \times a^7 \equiv a^4 \times a^7 \equiv a^{11} \equiv a \pmod{11}.$$

Working modulo 17:

$$a^{561} = (a^{17})^{33} \equiv a^{33} = a^{17} \times a^{16} \equiv a \times a^{16} \equiv a^{17} \equiv a \pmod{17}.$$

Theorem 1.3 of *Unit 3* now gives $a^{561} \equiv a \pmod{561}$ for all integers a. So 561 is a counterexample to the converse of FLT. ♦

It has been proved, in 1994, that there are in fact infinitely many numbers which provide a counterexample to the converse of FLT.

Having two seemingly different formulations of FLT we have to decide which one to use in a given situation. When confronted with a result about *all* integers a, as in Example 1.4, you must use $a^p \equiv a \pmod{p}$, to ensure that the argument caters for the case $a \equiv 0 \pmod{p}$. On the other hand, when dealing with a result concerning integers a for which $a \not\equiv 0 \pmod{p}$, as in Example 1.3, although $a^p \equiv a \pmod{p}$ still applies, the alternative congruence $a^{p-1} \equiv 1 \pmod{p}$ invariably leads to simpler computations.

Problem 1.4

Prove that 168 divides $a^6 - 1$, where a is any integer for which $\gcd(a, 42) = 1$.

Problem 1.5

For any integer a show that a^5 and a have the same units digit. Deduce that a^{100} has the same units digit as a^4.

2 REPRESENTATION OF FRACTIONS BY DECIMALS

2.1 Terminating and cycling decimals

When reciprocals of integers are written out as decimals a variety of patterns emerge. A few reciprocals are shown in Table 2.1.

Table 2.1 The decimal representation of the reciprocals of 2 to 13

Reciprocal	Decimal representation
$\frac{1}{2}$	0.5
$\frac{1}{3}$	$0.333\ldots = 0.\langle 3\rangle$
$\frac{1}{4}$	0.25
$\frac{1}{5}$	0.2
$\frac{1}{6}$	$0.1666\ldots = 0.1\langle 6\rangle$
$\frac{1}{7}$	$0.142857142857142857\ldots = 0.\langle 142857\rangle$
$\frac{1}{8}$	0.125
$\frac{1}{9}$	$0.111\ldots = 0.\langle 1\rangle$
$\frac{1}{10}$	0.1
$\frac{1}{11}$	$0.090909\ldots = 0.\langle 09\rangle$
$\frac{1}{12}$	$0.08333\ldots = 0.08\langle 3\rangle$
$\frac{1}{13}$	$0.076923076923076923\ldots = 0.\langle 076923\rangle$

The 'angle bracket' notation is used to indicate that the string of digits in the brackets repeats indefinitely.

We observe that some of the decimals terminate while others go on forever. All the non-terminating ones have a 'cycle' which is eventually repeated indefinitely, such as the cycle $\langle 142857\rangle$ of length 6 in $\frac{1}{7}$ and the cycle $\langle 09\rangle$ of length 2 in $\frac{1}{11}$. In some of the decimals (for example $\frac{1}{3}$ and $\frac{1}{7}$) there is just the repeating cycle, while in others (for example $\frac{1}{6}$ and $\frac{1}{12}$) the cycle is preceded by some dissimilar digits. Can we explain these differences?

The length of a cycle is the number of digits forming the cycle.

To determine the decimal of $\frac{1}{7}$ by hand you might well proceed by long division, as illustrated on the left in Figure 2.1 . To remind you of how long division works, at the boxed stage, 20 is divided by 7 giving a quotient of 2 and a remainder of 6. The remainder is then multiplied by 10 and the process continues by dividing 60 by 7. On the right we have presented the same set of calculations expressed in terms of the division algorithm.

From the system of equations we observe that the remainders which arise are the least positive residues modulo 7 of successive powers of 10. The seventh equation reveals that $10^6 \equiv 1 \pmod 7$, a fact that we could have predicted by virtue of FLT. Observe also that at the seventh equation we meet the first repeated remainder, as 1 was also the remainder in the original equation. The relevance of this is that thereafter the same sequence of equations must recur: the eighth equation is the second one again, and so on, explaining the cycling of the decimal.

There is nothing special about 7 here. For any positive integer n, as there are only n residues modulo n, a remainder must repeat, and from this point onwards the sequence of equations recurs.

Problem 2.1 ⎯⎯⎯⎯⎯⎯⎯⎯⎯⎯⎯⎯⎯⎯⎯⎯⎯⎯⎯⎯⎯⎯⎯⎯⎯⎯⎯

Determine, by using the division algorithm, the decimal of (a) $\frac{1}{15}$ and (b) $\frac{1}{41}$.

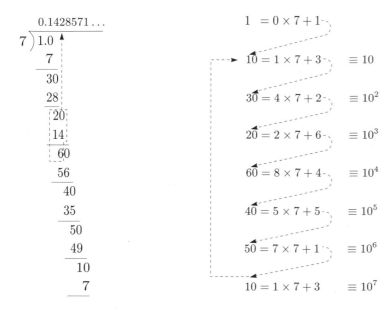

Notice that the zero quotient in the first equation just gives the zero before the decimal point.

Figure 2.1   The decimal of $\frac{1}{7}$ by long division

Let us look at the general reciprocal $\dfrac{1}{n}$ for the integer $n > 1$. Suppose that the equations on dividing 1 by n begin

$$
\begin{aligned}
1 &= 0n + 1 \\
10 &= q_1 n + r_1 & 0 \le r_1 < n \\
10 r_1 &= q_2 n + r_2 & 0 \le r_2 < n \\
10 r_2 &= q_3 n + r_3 & 0 \le r_3 < n \\
&\ \ \vdots
\end{aligned}
$$

Note that each q_i is one of the digits 0, 1, 2, ..., or 9 since $0 \le 10 r_{i-1} < 10n$, and q_i is the quotient on dividing $10 r_{i-1}$ by n.

Note also that, for each k,

$$r_k \equiv 10^k \pmod{n}.$$

The decimal of $\dfrac{1}{n}$ then begins $0.q_1 q_2 q_3 \ldots$. But it is the remainders r_i which tell us when recurrence will take place. As each of the remainders r_i is one of the n integers satisfying $0 \le r_i < n$, there are at most n different equations here, and once a remainder appears for the second time subsequent equations repeat in a cycle.

One possibility which we have not considered is that remainder 0 arises. If this happens the sequence of equations becomes, for some k

$$
\begin{aligned}
&\ \ \vdots \\
10 r_{k-2} &= q_{k-1} n + r_{k-1} \\
10 r_{k-1} &= q_k n + 0 \\
0 &= 0n + 0 \\
&\ \ \vdots
\end{aligned}
$$

This is the situation corresponding to the terminating decimal. The last non-zero equation tells us that n divides $10 r_{k-1}$, from which we deduce that n divides 10^k. Now the divisors of powers of 10 are those numbers whose only prime divisors are 2 and/or 5, namely the numbers of the form $2^r 5^s$, where $r \ge 0$ and $s \ge 0$. Conversely if $n = 2^r 5^s$, then n divides 10^k where $k = \max\{r, s\}$ and the decimal of $\dfrac{1}{n}$ terminates after k decimal places. So the integers $1 < n < 100$ for which the decimal of $\dfrac{1}{n}$ terminates are $n = 2, 4,$ 5, 8, 10, 16, 20, 25, 32, 40, 50, 64 and 80.

$r_{k-1} \equiv 10^{k-1} \pmod{n}$

We summarize these findings in the following result.

Theorem 2.1 Terminating decimals

The decimal of $\frac{1}{n}$ terminates if, and only if, $n = 2^r 5^s$, for some integers $r \geq 0$ and $s \geq 0$.

Diversion

$$1 \times 7 + 3 = 10$$
$$14 \times 7 + 2 = 100$$
$$142 \times 7 + 6 = 1000$$
$$1428 \times 7 + 4 = 10000$$
$$14285 \times 7 + 5 = 100000$$
$$142857 \times 7 + 1 = 1000000$$
$$1428571 \times 7 + 3 = 10000000$$
$$14285714 \times 7 + 2 = 100000000$$
$$142857142 \times 7 + 6 = 1000000000$$
$$1428571428 \times 7 + 4 = 10000000000$$
$$14285714285 \times 7 + 5 = 100000000000$$
$$142857142857 \times 7 + 1 = 1000000000000$$

2.2 The order of an integer

Let us turn to $\frac{1}{p}$ for p a prime other than 2 or 5. As $\gcd(p, 10) = 1$, FLT tells us that $10^{p-1} \equiv 1 \pmod{p}$. The implication of this for the system of equations for $\frac{1}{p}$ is that the remainder in the pth equation is 1, and this will be a repeat of the remainder in the initial equation.

$$1 = 0 \times p + 1$$
$$\vdots$$
$$10^{p-1} \equiv 1 = 0 \times p + 1$$

So the decimal of $\frac{1}{p}$ has a cycle which begins immediately after the decimal point and which has length at most $p - 1$. We say that the length is *at most* $p - 1$ because, although we know that $10^{p-1} \equiv 1 \pmod{p}$, it might also be the case that $10^k \equiv 1 \pmod{p}$ for some smaller positive integer k, and the remainder 1 would be repeated in an earlier equation. Indeed we have already seen that this can happen. When $p = 11$ the decimal of $\frac{1}{11}$ is $0.\langle 09 \rangle$. As predicted, this has a cycle beginning at the first decimal place. But the length of this cycle is 2 rather than the suggested 10, because it is the case that $10^2 \equiv 1 \pmod{11}$. Similarly, in Problem 2.1 we discovered that $\frac{1}{41}$ has a cycle of length 5. While FLT assured us that the cycle length could not exceed 40, since $10^{40} \equiv 1 \pmod{41}$, it is readily checked that $10^5 \equiv 1 \pmod{41}$.

We should really say that the decimal of $\frac{1}{41}$ has a cycle of length 5. However, when the meaning is clear we shall often omit 'the decimal of' for brevity.

All this suggests that we should look with interest at increasing positive powers of 10 searching for the first one which is congruent modulo p to 1, for this least exponent will give the length of the cycle in $\frac{1}{p}$. We refer to this least exponent as the *order* of 10 modulo p. The general definition is as follows.

Definition 2.1 The order of an integer

If p is prime and $\gcd(a, p) = 1$, then the *order* of a modulo p is the least positive integer c such that $a^c \equiv 1 \pmod{p}$.

If $\gcd(a, p) \neq 1$ then $a \equiv 0 \pmod{p}$ and so $a^c \equiv 0 \pmod{p}$ for all positive integers c. In this case there is no concept of order.

This definition assumes that for $\gcd(a, p) = 1$ there is a unique least positive integer c such that $a^c \equiv 1 \pmod{p}$. This is guaranteed by FLT since $a^{p-1} \equiv 1 \pmod{p}$ gives $c \leq p - 1$.

With this terminology at our disposal, we can state formally what we have discovered about the cycle in $\dfrac{1}{p}$.

Theorem 2.2 The cycle length of $\dfrac{1}{p}$

The length of the cycle in the decimal of $\dfrac{1}{p}$ is equal to the order of 10 modulo p, where p is a prime other than 2 or 5.

Problem 2.2 _____

Determine the order of 10 modulo 17 and hence the length of the cycle of $\frac{1}{17}$.

_____

Table 2.2 shows the order of 10 modulo each of the primes $p < 100$ (other than 2 and 5). Notice that for nine of these primes the order of 10 modulo p is $p - 1$. When this occurs we say that 10 is a *primitive root* of p, and in this case the successive powers of 10 run through all the $p - 1$ non-zero residues of p in some order, as illustrated for the prime 17 in Problem 2.2.

We deal with primitive roots in general in the next unit. For our present purposes we are concerned only with cases where 10 is a primitive root of p.

Table 2.2 The order of 10 modulo p for primes $p < 100$, other than 2 or 5.

Prime p	Order of 10 modulo p	Prime p	Order of 10 modulo p
3	1	47	46
7	6	53	13
11	2	59	58
13	6	61	60
17	16	67	33
19	18	71	35
23	22	73	8
29	28	79	13
31	15	83	41
37	3	89	44
41	5	97	96
43	21		

This is the start of a much more extensive table which illustrates how, over the years, mathematicians have devoted large portions of their lives to calculations which can only be their own reward. In 1873 William Shanks published an extension of this table covering all primes up to 30 000 and subsequently progressed the table to all primes less than 120 000. To get some idea of how much calculation this involved, and remember that we are talking about the days long before electronic calculating devices had landed on our desks, consider the prime 61. We know that $10^{60} \equiv 1 \pmod{61}$ and we need to confirm that there is no smaller positive integer k for which $10^k \equiv 1 \pmod{61}$. In Problem 2.2 we attacked a similar problem for the prime 17 by working out all the smaller powers of 10. There was no need to work them all out as the following illustrates.

Problem 2.3 _____

Prove that if the order of 10 modulo 61 is c, then c is a divisor of 60.

_____

By virtue of Problem 2.3 the order of 10 modulo 61 can be determined by finding the least positive residues, modulo 61, of the following powers of 10:

$$10, 10^2, 10^3, 10^4, 10^5, 10^6, 10^{10}, 10^{12}, 10^{15}, 10^{20} \text{ and } 10^{30}.$$

The required order is the exponent in the first of these which is congruent modulo 61 to 1. If, as turns out to be the case, none of them is congruent to 1 then the order of 10 is 60. This is a considerable improvement on having to determine all the first 60 powers of 10 but it still represents a considerable amount of computation; and all this is for just one relatively small prime. No doubt Shanks had a number of short-cuts available, but to achieve what he did by hand was no mean feat.

The result of Problem 2.3 can be generalized, for there is nothing special about the prime 61. The proof of the following result, which we shall use in the next unit, simply follows the argument given in the solution of Problem 2.3.

Theorem 2.3

If a has order c modulo p, where $p \geq 3$ is prime, and $a^k \equiv 1 \pmod{p}$ then c is a divisor of k.

In particular, c divides $p - 1$.

Proof of Theorem 2.3

By the Division Algorithm, we can write

$$k = qc + r, \quad \text{where } 0 \leq r < c.$$

Now

$$a^k = a^{qc+r} = (a^c)^q \times a^r \equiv 1^q \times a^r \equiv a^r \pmod{p}$$

and as $a^k \equiv 1 \pmod{p}$, $a^r \equiv 1 \pmod{p}$. Therefore, $r = 0$, for otherwise the order of a would be less than c. Hence c divides k.

As FLT gives $a^{p-1} \equiv 1 \pmod{p}$, the order c divides $p - 1$. ∎

2.3 Decimal representations of rational numbers

To determine the decimal of $\frac{2}{7}$ we proceed just as for $\frac{1}{7}$, but this time the first equation involves dividing 2 by 7.

$$
\begin{aligned}
2 &= 0 \times 7 + 2 \\
20 &= 2 \times 7 + 6 \\
60 &= 8 \times 7 + 4 \\
40 &= 5 \times 7 + 5 \\
50 &= 7 \times 7 + 1 \\
10 &= 1 \times 7 + 3 \\
30 &= 4 \times 7 + 2
\end{aligned}
$$

Notice that we get the same cycle of six equations as we had for $\frac{1}{7}$ but starting with a different equation. So the same six digits, in the same cyclic order but this time starting with the digit 2, form the cycle $\langle 285714 \rangle$ of $\frac{2}{7}$. In the same way you could check that $\frac{3}{7}$, $\frac{4}{7}$, $\frac{5}{7}$ and $\frac{6}{7}$ have decimals consisting of the same cycle of digits as that of $\frac{1}{7}$ but with appropriate starting points as shown below.

$$\frac{1}{7} = 0.\langle 142857 \rangle$$
$$\frac{2}{7} = 0.\langle 285714 \rangle$$
$$\frac{3}{7} = 0.\langle 428571 \rangle$$
$$\frac{4}{7} = 0.\langle 571428 \rangle$$
$$\frac{5}{7} = 0.\langle 714285 \rangle$$
$$\frac{6}{7} = 0.\langle 857142 \rangle$$

This situation is similar for any prime p having 10 as a primitive root. For such a prime the cycle of $\frac{1}{p}$ has length $p - 1$ and each $\frac{r}{p}$, for $r = 1, 2, \ldots,$ $p - 1$, has the same cycle of $p - 1$ digits but each with its own starting position.

Although the proof of this result is not too demanding we have opted not to include it here.

Problem 2.4

Given that $\frac{1}{17} = 0.\langle 0588235294117647 \rangle$ determine the decimals of $\frac{2}{17}$ and $\frac{9}{17}$.

We would be lacking in curiosity if we did not now ask whether or not any of this 'recycling' occurs when 10 is not a primitive root of prime p. Look at $p = 13$. In finding the decimal of $\frac{1}{13}$ we get the following set of equations.

$$1 = 0 \times 13 + 1$$
$$10 = 0 \times 13 + 10$$
$$100 = 7 \times 13 + 9$$
$$90 = 6 \times 13 + 12$$
$$120 = 9 \times 13 + 3$$
$$30 = 2 \times 13 + 4$$
$$40 = 3 \times 13 + 1$$

As the cycle has length six, the order of 10 modulo 13 is 6 and so 10 is not a primitive root of 13. By taking each of its six starting points these equations give following the six decimals:

$$\frac{1}{13} = 0.\langle 076923 \rangle, \quad \frac{3}{13} = 0.\langle 230769 \rangle, \quad \frac{4}{13} = 0.\langle 307692 \rangle,$$
$$\frac{9}{13} = 0.\langle 692307 \rangle, \quad \frac{10}{13} = 0.\langle 769230 \rangle, \quad \frac{12}{13} = 0.\langle 923076 \rangle.$$

In finding the decimal of $\frac{2}{13}$ we get the following set of equations.

The numerator in each fraction is the remainder in the equation which precedes the start of the cycle.

$$2 = 0 \times 13 + 2$$
$$20 = 1 \times 13 + 7$$
$$70 = 5 \times 13 + 5$$
$$50 = 3 \times 13 + 11$$
$$110 = 8 \times 13 + 6$$
$$60 = 4 \times 13 + 8$$
$$80 = 6 \times 13 + 2$$

These equations give a second set of six decimals:

$$\frac{2}{13} = 0.\langle 153846 \rangle, \quad \frac{5}{13} = 0.\langle 384615 \rangle, \quad \frac{6}{13} = 0.\langle 461538 \rangle,$$
$$\frac{7}{13} = 0.\langle 538461 \rangle, \quad \frac{8}{13} = 0.\langle 615384 \rangle, \quad \frac{11}{13} = 0.\langle 846153 \rangle.$$

There is nothing exceptional in this behaviour for the prime 13: it is typical of what happens for any prime p. Suppose that the order of 10 modulo p is c. It turns out that the decimals of $\frac{r}{p}$, for $1 \leq r < p$ form sets, each set consisting of c members with the same cycle of c digits varying only in their starting points. For example, consider $p = 41$. The order of 10 modulo 41 is 5 (see the solution of Problem 2.1) and the 40 decimals $\frac{r}{41}$ have cycles of five digits. So these decimals will form eight sets, each set consisting of the same cycle but with the five different starting points. These are displayed in Table 2.3.

Table 2.3 The sets of cycles of $\frac{r}{41}$, for $1 \leq r < 41$

Set			Decimals		
1	$\frac{1}{41} = 0.\langle 02439 \rangle$	$\frac{10}{41} = 0.\langle 24390 \rangle$	$\frac{16}{41} = 0.\langle 39024 \rangle$	$\frac{18}{41} = 0.\langle 43902 \rangle$	$\frac{37}{41} = 0.\langle 90243 \rangle$
2	$\frac{2}{41} = 0.\langle 04878 \rangle$	$\frac{20}{41} = 0.\langle 48780 \rangle$	$\frac{32}{41} = 0.\langle 78048 \rangle$	$\frac{33}{41} = 0.\langle 80487 \rangle$	$\frac{36}{41} = 0.\langle 87804 \rangle$
3	$\frac{3}{41} = 0.\langle 07317 \rangle$	$\frac{7}{41} = 0.\langle 17073 \rangle$	$\frac{13}{41} = 0.\langle 31707 \rangle$	$\frac{29}{41} = 0.\langle 70731 \rangle$	$\frac{30}{41} = 0.\langle 73170 \rangle$
4	$\frac{4}{41} = 0.\langle 09756 \rangle$	$\frac{23}{41} = 0.\langle 56097 \rangle$	$\frac{25}{41} = 0.\langle 60975 \rangle$	$\frac{31}{41} = 0.\langle 75609 \rangle$	$\frac{40}{41} = 0.\langle 97560 \rangle$
5	$\frac{5}{41} = 0.\langle 12195 \rangle$	$\frac{8}{41} = 0.\langle 19512 \rangle$	$\frac{9}{41} = 0.\langle 21951 \rangle$	$\frac{21}{41} = 0.\langle 51219 \rangle$	$\frac{39}{41} = 0.\langle 95121 \rangle$
6	$\frac{6}{41} = 0.\langle 14634 \rangle$	$\frac{14}{41} = 0.\langle 34146 \rangle$	$\frac{17}{41} = 0.\langle 41463 \rangle$	$\frac{19}{41} = 0.\langle 46341 \rangle$	$\frac{26}{51} = 0.\langle 63414 \rangle$
7	$\frac{11}{41} = 0.\langle 26829 \rangle$	$\frac{12}{41} = 0.\langle 29268 \rangle$	$\frac{28}{41} = 0.\langle 68292 \rangle$	$\frac{34}{41} = 0.\langle 82926 \rangle$	$\frac{38}{41} = 0.\langle 92682 \rangle$
8	$\frac{15}{41} = 0.\langle 36585 \rangle$	$\frac{22}{41} = 0.\langle 53658 \rangle$	$\frac{24}{41} = 0.\langle 58536 \rangle$	$\frac{27}{41} = 0.\langle 65853 \rangle$	$\frac{35}{41} = 0.\langle 85365 \rangle$

We shall leave decimals at this point, aware that there are further areas for investigation. In particular, we have not pursued reciprocals of composite numbers, other than to recognize which ones terminate. As a parting observation we mention that the decimal of $\frac{1}{49}$ is a cycle of length 42:

$$\frac{1}{49} = 0.\langle 020408163265306122448979591836734693877551 \rangle,$$

and that the forty two decimals obtained from this cycle by letting each digit of the cycle lead are the *proper* fractions $\frac{k}{49}$, where $1 \leq k < 49$; that is those with $\gcd(k, 49) = 1$. For example

$$\frac{11}{49} = 0.\langle 224489795918367346938775510204081632653061 \rangle.$$

A proper faction is one of the form $\frac{m}{n}$, where $\gcd(m, n) = 1$, i.e. a fraction in which all common divisors of the numerator and denominator have been cancelled.

However, of all composite integers less than 100, 49 is the only one which exhibits this elegant behaviour of its proper fractions. The relevance of the cycle length of $1/49$ will become apparent once we have met Euler's generalization of FLT, in the next unit.

3 WILSON'S THEOREM

Let us turn now to a result which is not of very much practical use but one which, nevertheless, has a prominent place in the development of number theory. The English mathematician Edward Waring reported, in his book *Meditationes Algebraicae* (1770), a result claimed by one of his former students, John Wilson. Wilson had asserted that for each prime p, the number

$$\frac{(p-1)! + 1}{p}$$

is an integer. Wilson appears to have guessed the result on empirical evidence for neither he nor Waring could supply a proof. But shortly afterwards the French mathematician Lagrange gave a proof of the result. Throughout mathematical literature the result has become known as Wilson's Theorem, although a manuscript written in 1682 indicates that Leibnitz was aware of the result, and how to prove it, long before Wilson came on the scene.

The language of congruence gives us a neater way of expressing the result.

Theorem 3.1 Wilson's Theorem

If p is prime then $(p-1)! \equiv -1 \pmod{p}$.

We shall first look at the particular case of the prime 17 to illustrate the idea behind the proof that we are going to give. For $p = 17$ the theorem claims that $16! \equiv -1 \pmod{17}$.

To evaluate $16!$ we could multiply together the sixteen numbers from 1 to 16 inclusive, reducing the answer modulo 17. But notice how the numbers from 2 to 15 inclusive pair up so that their product is congruent modulo 17 to 1.

$$2 \times 9 \equiv 1 \pmod{17} \qquad 3 \times 6 \equiv 1 \pmod{17}$$
$$4 \times 13 \equiv 1 \pmod{17} \qquad 5 \times 7 \equiv 1 \pmod{17}$$
$$8 \times 15 \equiv 1 \pmod{17} \qquad 10 \times 12 \equiv 1 \pmod{17}$$
$$11 \times 14 \equiv 1 \pmod{17}$$

This observation gives us that

$$15! = 2 \times 3 \times 4 \times 5 \times 6 \times 7 \times 8 \times 9 \times 10 \times 11 \times 12 \times 13 \times 14 \times 15$$
$$\equiv 1^7 \equiv 1 \pmod{17},$$

and so

$$16! = 15! \times 16 \equiv 1 \times 16 \equiv 16 \equiv -1 \pmod{17}.$$

The substance of our general proof lies in showing that this convenient pairing of divisors in $(p-2)!$ always occurs. This particular proof was first given by Gauss.

Proof of Wilson's Theorem (Theorem 3.1)

The observations $(2-1)! = 1 \equiv -1 \pmod{2}$ and $(3-1)! = 2 \equiv -1 \pmod{3}$ confirm the result for $p = 2$ and $p = 3$, so we can confine attention to $p \geq 5$.

Consider the set of $p-3$ integers $S = \{2, 3, 4, \ldots, p-2\}$. If $a \in S$ then, as $\gcd(a, p) = 1$, the linear congruence $ax \equiv 1 \pmod{p}$ has, (by Theorem 3.2 of *Unit 3*), a unique least positive solution modulo p. Let the solution be a', so that $aa' \equiv 1 \pmod{p}$. Now $a' \neq 1$ for that would give $a \equiv 1 \pmod{p}$, contradicting $a \in S$. Similarly $a' \neq p-1$ for that would imply $a \equiv -1 \pmod{p}$, again contradicting $a \in S$. Hence $a' \in S$.

Finally $a \neq a'$, for otherwise $aa' \equiv a^2 \equiv 1 \pmod{p}$, which would give that p divides $a^2 - 1$. But then p divides $(a-1)(a+1)$ whereupon Euclid's Lemma gives p divides $a-1$ or p divides $a+1$ which amount to $a = 1$ or $a = p-1$ respectively, each of which contradicts $a \in S$.

We conclude that for each element $a \in S$ there is a unique element $a' \in S$, $a \neq a'$, for which $aa' \equiv 1 \pmod{p}$. Thus the elements of S form $\dfrac{p-3}{2}$ distinct pairs a and a' with $aa' \equiv 1 \pmod{p}$. Multiplying these congruences together, each element of S is involved exactly once, and so

Note that $p - 3$ is even as p is odd.

$$2 \times 3 \times 4 \times \cdots \times (p-2) \equiv 1 \pmod{p}.$$

Multiplying both sides of this congruence by $(p-1)$ gives

$$2 \times 3 \times 4 \times \cdots \times (p-2) \times (p-1) \equiv p-1 \pmod{p}.$$

That is,

$$(p-1)! \equiv -1 \pmod{p},$$

completing the proof. ∎

The converse of Wilson's Theorem is also true, and its proof is relatively simple.

Theorem 3.2 Converse of Wilson's Theorem

If $n > 1$ is an integer and $(n - 1)! \equiv -1 \pmod{n}$ then n is prime.

Proof of Theorem 3.2

Suppose that n divides $(n - 1)! + 1$. Let r be a positive divisor of n, where $r < n$. Then as

$$(n - 1)! = 1 \times 2 \times 3 \times \cdots \times r \times \cdots \times (n - 1),$$

r divides $(n - 1)!$. But r divides $(n - 1)! + 1$, and so we are forced to conclude that $r = 1$. That is, n has no positive divisors other than itself and 1, and therefore is prime. ∎

Problem 3.1

Find the smallest prime divisor of

(a) $18! + 1$, and (b) $29! - 1$.

Wilson's Theorem together with its converse give a characterization of prime numbers: $n > 1$ is prime if, and only if, n divides $(n - 1)! + 1$. Unfortunately it is not of much practical use because of the amount of computation that would be involved in using it. For instance, to prove that 43 is prime by this method involves checking whether or not $42! + 1$ is divisible by 43. Even working modulo 43 the evaluation of $42!$ is quite demanding, and that is for a relatively small prime. However, Wilson's Theorem can be put to work in other ways. In the next example it is used to discover a solution of a *quadratic congruence*. The theory of quadratic congruences has great importance in number theory and we shall devote a substantial part of *Unit 6* to its study.

Example 3.1

Show that the congruence $x^2 + 1 \equiv 0 \pmod{p}$, where p is an odd prime, has a solution if, and only if, $p \equiv 1 \pmod{4}$.

We first use Wilson's Theorem to show that the congruence has a solution when $p = 4k + 1$. In this case $(p - 1)! = (4k)! \equiv -1 \pmod{4k + 1}$. That is,

$$(p - 1)! = 1 \times 2 \times 3 \times \cdots \times (2k) \times (2k + 1) \times \cdots \times (4k - 1) \times (4k)$$
$$\equiv -1 \pmod{4k + 1}.$$

Now working modulo $4k + 1$ we have

$$4k \equiv -1$$
$$4k - 1 \equiv -2$$
$$\vdots$$
$$2k + 1 \equiv -2k,$$

and substituting these values in the expression for $(p - 1)!$ we obtain

$$-1 \equiv (p - 1)! \pmod{4k + 1}$$
$$\equiv 1 \times 2 \times \cdots \times (2k) \times (-2k) \times \cdots \times (-2) \times (-1) \pmod{4k + 1}$$
$$\equiv 1 \times 2 \times \cdots \times (2k) \times (-1)^{2k} \times (2k) \times \cdots \times 2 \times 1 \pmod{4k + 1}$$
$$\equiv 1^2 \times 2^2 \times \cdots \times (2k)^2 \pmod{4k + 1}$$
$$\equiv ((2k)!)^2 \pmod{4k + 1}.$$

Hence $x = (2k)!$ satisfies $x^2 + 1 \equiv 0 \pmod{4k + 1}$ and so the congruence has a solution.

We now show that if $p \not\equiv 1 \pmod 4$ then there is no solution. In this case the odd prime is of the form $4k + 3$. Heading for a contradiction, suppose that the congruence $x^2 + 1 \equiv 0 \pmod{4k+3}$ has a solution; that is, there is an integer a such that $a^2 \equiv -1 \pmod{4k+3}$. Then, applying FLT for prime $4k + 3$:

$$a^{4k+2} \equiv 1 \pmod{4k+3}.$$

But

$$a^{4k+2} = (a^2)^{2k+1} \equiv (-1)^{2k+1} \equiv -1 \pmod{4k+3},$$

giving the contradiction $1 \equiv -1 \pmod{4k+3}$. ◆

Problem 3.2

Find two solutions of the quadratic congruence $x^2 + 1 \equiv 0 \pmod{29}$.

Problem 3.3

Prove that, for each odd prime p,

$$1^2 \times 3^2 \times \cdots \times (p-2)^2 \equiv 2^2 \times 4^2 \times \cdots \times (p-1)^2 \equiv (-1)^{(p+1)/2} \pmod p.$$

Hint: Use the fact that $2k \equiv -(p - 2k) \pmod p$.

As a final deduction from Wilson's Theorem we observe that it tells us that there are infinitely many integers n for which $n! + 1$ is composite: simply choose $n = p - 1$ for any prime $p > 3$. For other values of n the number $n! + 1$ may be either prime or composite, and it remains an unanswered question as to whether or not there are infinitely many primes of the form $n! + 1$. For values of n up to 100, those for which $n! + 1$ is prime are $n = 1$, 2, 3, 11, 27, 37, 41, 73 and 77.

As p is known to be a divisor of $(p - 1)! + 1$, the latter is composite except when $(p - 1)! + 1 = p$. This condition occurs only for $p = 2$ and $p = 3$.

4 POLYNOMIAL CONGRUENCES

4.1 Lagrange's Theorem

In preparation for the investigation of linear congruences in *Unit 3*, we briefly introduced the notion of polynomial congruences and their solutions. For convenience we reiterate the definitions here.

Definition 4.1 Polynomial congruences and their solutions

A polynomial congruence is an expression

$$P(x) = c_r x^r + c_{r-1} x^{r-1} + \cdots + c_1 x + c_0 \equiv 0 \pmod n,$$

where $P(x)$ is a polynomial of degree $r \geq 0$ with integer coefficients.

An integer a is a solution of the polynomial congruence $P(x) \equiv 0 \pmod n$ if, and only if, $P(a) \equiv 0 \pmod n$.

The number of solutions of a polynomial congruence is the number of incongruent solutions modulo n.

In the congruence context we are concerned exclusively with integers and consequently we are forced to restrict attention to polynomials with integer coefficients. Hence, in this section, when we mention a polynomial it will be assumed that it has integer coefficients.

It follows from these definitions that the polynomial congruence $P(x) \equiv 0 \pmod{n}$ has at most n solutions, and these could be discovered by finding which integers in a complete set of residues, such as $\{0, 1, 2, \ldots, n-1\}$, are solutions.

Problem 4.1 _____

Solve the following polynomial congruences.

(a) $x^3 - x \equiv 0 \pmod{6}$

(b) $x^3 - x \equiv 0 \pmod{8}$

(c) $x^2 + x + 1 \equiv 0 \pmod{7}$

(d) $2x^2 + 9x + 15 \equiv 0 \pmod{7}$

_____

Consider the polynomial congruence

$$x^{20} + 3x^{14} + 8x^{10} + 3x^2 + 6 \equiv 0 \pmod{7}.$$

This still may not appear too daunting. After all we could solve the congruence by exhaustion, evaluating the left-hand side for $x = 0$, 1, 2, 3, 4, 5 and 6. But even that involves a good deal of computation, and if we change the modulus to something significantly larger this approach would be impracticable. What we need is a more sophisticated, systematic approach, and here FLT comes to our aid.

FLT informs us that, for all integers x, $x^7 \equiv x \pmod{7}$, and so

$$x^{20} = (x^7)^2 x^6 \equiv x^2 x^6 \equiv x^7 x \equiv x^2 \pmod{7}.$$

Similarly, $x^{14} \equiv x^2 \pmod{7}$ and $x^{10} \equiv x^4 \pmod{7}$. Therefore solving the original congruence is equivalent to solving

$$x^2 + 3x^2 + 8x^4 + 3x^2 + 6 = 8x^4 + 7x^2 + 6 \equiv 0 \pmod{7}.$$

Simplifying further by reducing the coefficients, $8 \equiv 1 \pmod{7}$, etc, we arrive at the equivalent congruence

$$x^4 - 1 \equiv 0 \pmod{7}.$$

Checking the fourth powers of 0, ± 1, ± 2 and ± 3, we find that the congruence has just the two solutions $x \equiv \pm 1 \pmod{7}$.

This example illustrates a useful role for FLT in solving polynomial congruences, offering the possibility of replacing the polynomial in question by one of smaller degree. We are not going to develop any general theory for solving such congruences here but, rather, just touch the beginnings of the subject through examples. What we shall find is that when working to a modulus which is not prime, unexpected and unwanted things can happen; but when the modulus is prime the situation is more clear-cut. Our principal goal is a theorem of Lagrange which gives an insight into the number of solutions to a polynomial congruence with prime modulus.

Joseph Louis Lagrange (1736–1813)

Lagrange was born in Turin to a French father and Italian mother, but he spent the most productive years of his life in Germany. His parents both came from wealthy backgrounds. However his father was an unsuccessful gambler with the result that Lagrange inherited very little. This could well have been fortunate for mathematics, for had Lagrange been wealthy he would, by his own admission, have been lost to mathematics.

In his early education Lagrange was interested in the classics. But on encountering an essay by Halley — Newton's friend, made famous by the comet named after him — extolling the relative merits of calculus over traditional geometric approaches to problems, he was converted to mathematics, and by the time he was 18 years old he was appointed

Professor of Mathematics at the Royal Artillery School in Turin. This was the beginning of a great career.

Lagrange was an analyst, not a geometer. His early researches centred on applications of calculus and culminated in his masterpiece the *Mecanique Analytique* (Analytical Mechanics). Pushing his view that mechanics was really a branch of pure mathematics in which geometric ideas played no part, he remarks in the preface that the science of mechanics can be considered to be a geometry of a space of four dimensions — three Cartesian coordinates and one time coordinate. Despite this view, it is interesting to note that there are no diagrams in this work.

Through private communications Euler recognized a great talent in the young Lagrange. Lagrange sent Euler a solution to the *isoperimetric problem* (a famous problem in the calculus of variations) using his own methods. This was a problem which Euler admitted had baffled him for many years. In 1766 when he was due to leave the Berlin Academy for St Petersburg, Euler schemed successfully to get Lagrange installed as his replacement. A message from Frederick the Great invited 'the greatest mathematician in Europe' to live near 'the greatest king in Europe'. Lagrange accepted, and worked at Berlin for 20 years until, on the death of Frederick in 1787, he settled in Paris and took up French citizenship.

For the next six years Lagrange contributed nothing to mathematics. The many years of excessively hard work had taken their toll and his appetite for mathematics had gone. He suffered long fits of depression. Strangely, it was the turmoil of the French Revolution which revitalized his interest in mathematics. Following the abolition of the old universities, the revolutionists created the Ecole Normale in 1795, and Lagrange was its first Professor of Mathematics. When the Normale closed in 1797 Lagrange moved on to become professor at the Ecole Polytechnique. During this period Lagrange gained a reputation as an outstanding teacher. His courses of lectures, and the recognition of the difficulties his students had with some of the concepts, led to his writing two classic works, the *Theory of Analytic Functions* and *Lessons on the Calculus of Functions*.

Although Lagrange's main researches were in the domains of calculus and function theory, his interests were much more general and, like so many other greats before and after him, he had a special talent for number theory. His name is associated with solutions of many Diophantine equations (which we shall meet in *Unit 8*) and, in particular, Lagrange gave the first proof that every positive integer can be expressed as the sum of four squares. In this unit we shall see a result of Lagrange which is concerned with solutions of polynomial congruences.

One well-known principle from the theory of equations, the so-called Fundamental Theorem of Algebra, asserts that a polynomial equation of degree n with real coefficients has at the most n roots (solutions). We already have evidence that the same principle cannot hold for polynomial congruences; for example, we witnessed in Problem 4.1 a polynomial of degree 3 for which the associated congruence modulo 6 has more than three solutions. Indeed, we have seen that linear congruences (degree 1) do not necessarily have unique solutions; the congruence $2x \equiv 2 \pmod{4}$ has the two solutions $x \equiv 1, 3 \pmod{4}$. However, looking back through the examples at hand, wherever the modulus is a prime this 'misbehaviour' does not occur. Lagrange, in 1768, proved that the number of solutions of a polynomial congruence in which the modulus is prime, is less than or equal to its degree.

But first, we must clarify what we mean by the degree of a polynomial congruence. Consider the following congruence.

$$7x^3 + 4x^2 - 3 \equiv 0 \ (\text{mod } 7).$$

The polynomial involved has degree 3. However, when we turn to congruence modulo 7, the leading term, $7x^3$, vanishes since $7 \equiv 0 \ (\text{mod } 7)$. Hence modulo 7, this polynomial has degree 2 because the leading term with non-zero coefficient modulo 7 is $4x^2$.

By the *leading term* in a polynomial $P(x)$ we mean the (non-zero) term involving the highest power of x. By 'non-zero modulo n' we mean 'not congruent to zero modulo n'.

Definition 4.2 Degree of a polynomial congruence

If $P(x) = c_r x^r + c_{r-1}x^{r-1} + \cdots + c_1 x + c_0$ then the polynomial $P(x)$ has degree k modulo n if, after removing each term in which the coefficient $c_i \equiv 0 \ (\text{mod } n)$, the leading term remaining is $c_k x^k$.

If every coefficient $c_i \equiv 0 \ (\text{mod } n)$ then the polynomial is not assigned a degree modulo n.

The degree of the polynomial congruence $P(x) \equiv 0 \ (\text{mod } n)$ is the degree of $P(x)$ modulo n.

Problem 4.2 _____

What are the degrees of the following polynomial congruences?

(a) $3x^6 + 7x^2 - 2 \equiv 0 \ (\text{mod } 5)$

(b) $10x^3 + 3x + 1 \equiv 0 \ (\text{mod } 5)$

(c) $15x^2 - 10x + 2 \equiv 0 \ (\text{mod } 5)$

We are now ready to state our main result of this section. Its proof will come in the next subsection.

Theorem 4.1 Lagrange's Theorem

Let p be a prime. A polynomial congruence $P(x) \equiv 0 \ (\text{mod } p)$ of degree $k \geq 1$ can have at most k solutions.

4.2 Factorizing polynomial congruences

The proof that we shall give for Lagrange's Theorem follows closely one of the methods of proving the Fundamental Theorem of Algebra in which the polynomial $P(x)$ is effectively factorized. To this end we shall first prove a subsidiary result which introduces the idea of factors in polynomial congruences.

Theorem 4.2

Let b be a solution of the polynomial congruence $P(x) \equiv 0 \ (\text{mod } p)$ of degree $k \geq 1$. Then $P(x) \equiv (x - b)P_1(x) \ (\text{mod } p)$, where $P_1(x)$ is a polynomial of degree $k - 1$ modulo p.

Proof of Theorem 4.2

Suppose that $P(x) = c_k x^k + c_{k-1} x^{k-1} + \cdots + c_1 x + c_0$, where $c_k \not\equiv 0 \pmod{p}$. Since $P(b) \equiv 0 \pmod{p}$ we have $P(x) \equiv P(x) - P(b) \pmod{p}$. Now

$$P(x) - P(b) = c_k(x^k - b^k) + c_{k-1}(x^{k-1} - b^{k-1}) + \cdots + c_1(x - b).$$

Each term in the sum on the right-hand side has a factor $(x - b)$ since

$$x^r - b^r = (x - b)(x^{r-1} + bx^{r-2} + b^2 x^{r-3} + \cdots + b^{r-2} x + b^{r-1}).$$

Hence

$$P(x) \equiv P(x) - P(b) \equiv (x - b)P_1(x) \pmod{p},$$

where the polynomial $P_1(x)$ has integer coefficients and leading term $c_k x^{k-1}$. As $c_k \not\equiv 0 \pmod{p}$, $P_1(x)$ has degree $k - 1$ modulo p. ■

In the statement of Theorem 4.2 we called the modulus p, but we did not stipulate that p was prime. Indeed the above proof holds true for p composite. However, when we progress from a single factor to our next result, which pulls out a factor for each solution of the congruence, we do need the modulus to be prime.

Theorem 4.3 Factorizing a polynomial modulo p

Let p be prime and let $b_1, b_2, \ldots, b_r$ be incongruent solutions of $P(x) \equiv 0 \pmod{p}$ which has degree $k \geq r$. Then

$$P(x) \equiv (x - b_1)(x - b_2) \ldots (x - b_r)P_r(x) \pmod{p},$$

where the polynomial $P_r(x)$ has degree $k - r$ modulo p.

Proof of Theorem 4.3

Theorem 4.2 tells us that $P(x) \equiv (x - b_1)P_1(x) \pmod{p}$, where $P_1(x)$ has degree $k - 1$ modulo p. Now $P(b_2) \equiv 0 \pmod{p}$, and so

$$(b_2 - b_1)P_1(b_2) \equiv 0 \pmod{p}.$$

But p does not divide $b_2 - b_1$, since b_1 and b_2 are incongruent solutions modulo p. Hence $P_1(b_2) \equiv 0 \pmod{p}$ by Euclid's Lemma. This allows us to invoke Theorem 4.2 again:

This is where we require the modulus p to be prime.

$$P_1(x) \equiv (x - b_2)P_2(x) \pmod{p},$$

where $P_2(x)$ has degree $k - 2$ modulo p.

Substituting for $P_1(x)$ gives

$$P(x) \equiv (x - b_1)(x - b_2)P_2(x) \pmod{p}.$$

Continuing, from $P(b_3) \equiv 0 \pmod{p}$ we deduce that $P_2(b_3) \equiv 0 \pmod{p}$ and hence that

$$P_2(x) \equiv (x - b_3)P_3(x) \pmod{p},$$

where $P_3(x)$ has degree $k - 3$ modulo p.

The desired result is reached when all r incongruent solutions have been processed in this way. ■

Let us look at an example to illustrate the proof of Theorem 4.3.

Example 4.1

Let $P(x) = x^4 - x^3 + 5x^2 - 3x - 2$. Confirm that $x \equiv 1, 3 \pmod{11}$ are two solutions of $P(x) \equiv 0 \pmod{11}$ and find $P_2(x)$ such that $P(x) \equiv (x-1)(x-3)P_2(x) \pmod{11}$.

$$P(1) = 1 - 1 + 5 - 3 - 2 = 0 \pmod{11}$$

and

$$P(3) = 81 - 27 + 45 - 9 - 2 = 88 \equiv 0 \pmod{11},$$

so both are solutions as claimed.

Writing $x^4 - x^3 + 5x^2 - 3x - 2 \equiv (x-1)(Ax^3 + Bx^2 + Cx + D) \pmod{11}$ we can solve for A, B, C and D by looking at the various powers of x and equating coefficients modulo 11.

$$
\begin{aligned}
x^4: &\quad 1 \equiv A \\
x^3: &\quad -1 \equiv -A + B, \text{ giving } B \equiv 0 \\
x^2: &\quad 5 \equiv -B + C, \text{ giving } C \equiv 5 \\
x\;\;: &\quad -3 \equiv -C + D, \text{ giving } D \equiv 2 \\
\text{constant}: &\quad -2 \equiv -D, \text{ which provides a check.}
\end{aligned}
$$

So $x^4 - x^3 + 5x^2 - 3x - 2 \equiv (x-1)(x^3 + 5x + 2) \pmod{11}$.

Now $x \equiv 3 \pmod{11}$ is a solution of $x^3 + 5x + 2 \equiv 0 \pmod{11}$ and so

$$x^3 + 5x + 2 \equiv (x-3)(Ex^2 + Fx + G) \pmod{11}.$$

We can solve this in the same way as above to get $E \equiv 1$, $F \equiv 3$, $G \equiv 3 \pmod{11}$.

Hence

$$x^4 - x^3 + 5x^2 - 3x - 2 \equiv (x-1)(x-3)(x^2 + 3x + 3) \pmod{11}. \qquad \blacklozenge$$

In fact the quadratic $x^2 + 3x + 3$ cannot be factorized modulo 11. In just the same way as a quadratic equation might have 0, 1 or 2 real solutions, it turns out that a quadratic congruence with prime modulus can have 0, 1 or 2 solutions. The quadratic congruence

$$x^2 + 3x + 3 \equiv 0 \pmod{11}$$

has no solutions.

Problem 4.3 _____

For each of the following polynomials $P(x)$, find all the solutions of $P(x) \equiv 0 \pmod{7}$ by exhaustion, and hence factorize the polynomial modulo 7.

(a) $x^2 + 4x - 4$

(b) $x^3 + 2x^2 + x + 3$

_____

If we apply Theorem 4.3 to one particular polynomial we get a result which is well worth recording.

Corollary to Theorem 4.3

For any prime p

$$x^{p-1} - 1 \equiv (x-1)(x-2)(x-3)\ldots(x-(p-1)) \pmod{p}.$$

Proof of the Corollary

Consider the congruence

$$x^{p-1} - 1 \equiv 0 \pmod{p}.$$

FLT uncovers $p-1$ incongruent solutions since this congruence is satisfied by $x = 1, 2, 3, \ldots, p-1$. Hence, applying Theorem 4.3

$$x^{p-1} - 1 \equiv (x-1)(x-2)(x-3)\ldots(x-(p-1))Q(x) \pmod{p},$$

where $Q(x)$ has degree $p - 1 - (p-1) = 0$ modulo p. So $Q(x)$ is a constant polynomial. Inspection of the coefficient of x^{p-1} confirms that $Q(x) \equiv 1 \pmod{p}$, as required. $\blacksquare$

It would be remiss of us not to record what happens when x is taken to be 0 in the Corollary.

For any prime p,

$$0 - 1 \equiv (-1)(-2)\cdots(-(p-1)) \pmod{p}$$
$$\equiv (-1)^{p-1}(p-1)! \pmod{p}$$
$$\equiv (p-1)! \pmod{p}, \quad \text{since } p-1 \text{ is even or}$$
$$p = 2 \text{ in which case} - 1 \equiv 1 \pmod{p}.$$

That is,

$$(p-1)! \equiv -1 \pmod{p},$$

which is Wilson's Theorem.

The result of Theorem 4.3 is the backbone of our proof of Lagrange's Theorem. We simply have to draw the right conclusions from our ability to pull out a linear factor corresponding to each solution of the congruence.

Proof of Lagrange's Theorem

Suppose to the contrary that a polynomial congruence $P(x) \equiv 0 \pmod{p}$ has degree k but has incongruent solutions $b_1, b_2, \ldots, b_k$ and b_{k+1}. Applying Theorem 4.3 for the first k solutions we have that

$$P(x) \equiv (x - b_1)(x - b_2)\ldots(x - b_k)P_k(x) \pmod{p},$$

where $P_k(x)$ has degree $k - k$ modulo p. Having degree 0, $P_k(x)$ is a constant polynomial, say, $P_k(x) = a$. Notice that the leading term in $P(x)$ modulo p is ax^k and so, as $P(x)$ has degree k modulo p, $a \not\equiv 0 \pmod{p}$.

The constant a is just the coefficient c_k of Definition 4.2.

Now consider the further solution b_{k+1}:

$$P(b_{k+1}) \equiv (b_{k+1} - b_1)(b_{k+1} - b_2)\ldots(b_{k+1} - b_k)a \equiv 0 \pmod{p}.$$

This means that p divides $(b_{k+1} - b_1)(b_{k+1} - b_2)\ldots(b_{k+1} - b_k)a$ and so, p being prime, Euclid's Lemma implies that p divides one of the terms $b_{k+1} - b_i$ or p divides a. But neither of these is possible: p cannot divide $b_{k+1} - b_i$ as the solutions are incongruent modulo p, and we have seen that $a \not\equiv 0 \pmod{p}$.

What has been contradicted? The only assumption made is that the polynomial congruence of degree k has $k + 1$ solutions. This therefore cannot happen and our proof is complete. ■

It is illuminating to see what happens when the modulus is not prime. Consider again the polynomial congruence

$$x^3 - x \equiv 0 \pmod{6},$$

which we found, in Problem 4.1, to have six solutions. One factorization of the polynomial $x^3 - x$ is readily spotted:

$$x^3 - x = x(x^2 - 1) = x(x - 1)(x + 1) \equiv 0 \pmod{6}.$$

Each of the three discovered factors suggests a solution of the congruence: $x \equiv 0$, 1 and $-1 \pmod{6}$ respectively. But $x \equiv 2$, 3 and 4 $\pmod{6}$ are also solutions, and the factorization has not contradicted this. For example, putting $x = 4$ into the factorized polynomial gives

$$4(4 - 1)(4 + 1) = 60 \equiv 0 \pmod{6},$$

and so the solution $x \equiv 4 \pmod{6}$ has not gone away!

The problem is that, since 6 is composite, we cannot deduce from $x(x - 1)(x + 1) \equiv 0 \pmod{6}$ that $x \equiv 0 \pmod{6}$, or $(x - 1) \equiv 0 \pmod{6}$ or $(x + 1) \equiv 0 \pmod{6}$. The corresponding deduction in Lagrange's Theorem could be made because the modulus was prime, and this was the key to relating factors of the polynomial to solutions of the congruence.

Notice that since $x \equiv 4 \pmod 6$ is a solution of this congruence, Theorem 4.2 claims that $x - 4$ can be pulled out as a factor on the left-hand side of $x^3 - x \equiv 0 \pmod 6$. It can, showing that the one factorization which we wrote down above is not the only one!

$$x^3 - x \equiv (x - 4)(x^2 + 4x + 3) \pmod 6$$
$$\equiv (x - 4)(x + 1)(x + 3) \pmod 6.$$

There are other ways of writing $x^3 - x$ as a product of three linear factors to which it is congruent modulo 6.

To complete the theoretical results of the section we give an interesting consequence of Lagrange's Theorem.

Theorem 4.4

If p is prime and d is a divisor of $p - 1$ then the congruence

$$x^d - 1 \equiv 0 \pmod p$$

has exactly d solutions.

Proof of Theorem 4.4

Let $p - 1 = dr$ for some integer r. If $r = 1$ then $d = p - 1$ and the result follows from the corollary to Theorem 4.3. We may assume therefore that $r \geq 2$. Then

$$x^{p-1} - 1 = (x^d - 1)(x^{d(r-1)} + x^{d(r-2)} + \cdots + x^d + 1)$$
$$= (x^d - 1)Q(x),$$

where $Q(x)$ has degree $d(r - 1)$ modulo p.

Now we know from the corollary to Theorem 4.3 that $x^{p-1} - 1 \equiv 0 \pmod p$ has $p - 1$ solutions. If a is such a solution then

$$a^{p-1} - 1 = (a^d - 1)Q(a) \equiv 0 \pmod p,$$

from which Euclid's Lemma informs us that either $(a^d - 1) \equiv 0 \pmod p$ or $Q(a) \equiv 0 \pmod p$. In other words, any solution of $x^{p-1} - 1 \equiv 0 \pmod p$ is necessarily a solution of either $x^d - 1 \equiv 0 \pmod p$ or $Q(x) \equiv 0 \pmod p$. Hence these latter two congruences have, between them, at least $p - 1$ solutions.

Now apply Lagrange's Theorem. The congruence $x^d - 1 \equiv 0 \pmod p$ has at most d solutions and $Q(x) \equiv 0 \pmod p$ has at most $d(r - 1)$ solutions. So between them they have at most

$$d + d(r - 1) = dr = p - 1$$

solutions. It follows that they have, between them, exactly $p - 1$ solutions and, in particular, $x^d - 1 \equiv 0 \pmod p$ has exactly d solutions. ∎

Diversion

$$3^2 + 4^2 = 5^2$$
$$10^2 + 11^2 + 12^2 = 13^2 + 14^2$$
$$21^2 + 22^2 + 23^2 + 24^2 = 25^2 + 26^2 + 27^2$$
$$36^2 + 37^2 + 38^2 + 39^2 + 40^2 = 41^2 + 42^2 + 43^2 + 44^2$$
$$55^2 + 56^2 + 57^2 + 58^2 + 59^2 + 60^2 = 61^2 + 62^2 + 63^2 + 64^2 + 65^2$$

4.3 Congruences with composite moduli

This section has been mainly concerned with theoretical results concerning polynomial congruences rather than with developing techniques for actually solving the congruences. For our purposes in this course we shall not have a great need to solve anything other than relatively simple congruences. We have seen that for not too large a prime modulus we can, failing anything better, attack congruences by an exhaustive method involving evaluation of the polynomial over a complete set of residues. To round off the section we shall look at a couple of more complicated examples with composite moduli which illustrate general lines of attack.

Consider the polynomial congruence $P(x) \equiv 0 \pmod{m}$, where $m = p_1^{k_1} p_2^{k_2} \ldots p_r^{k_r}$. If a is a solution of this congruence then $P(a) \equiv 0 \pmod{m}$ and so $P(a) \equiv 0 \left(\mathrm{mod}\ p_i^{k_i} \right)$ for each of the prime divisors. It follows that any solution of $P(x) \equiv 0 \pmod{m}$ must be a simultaneous solution of the r polynomial congruences $P(x) \equiv 0 \left(\mathrm{mod}\ p_i^{k_i} \right)$. Conversely if a is a simultaneous solution of the r polynomial congruences then by repeated application of the corollary to Theorem 1.3 of *Unit 3*, a is a solution of $P(x) \equiv 0 \pmod{m}$. So if we solve, individually, each of the congruences involving a single prime, we can then use the Chinese Remainder Theorem to build a solution of the original congruence. Here is a simple example.

Example 4.2

Solve $2x^2 + 5x - 3 \equiv 0 \pmod{72}$.

As $72 = 2^3 \times 3^2$, we first solve $2x^2 + 5x - 3 \equiv 0 \pmod{8}$ and $2x^2 + 5x - 3 \equiv 0 \pmod{9}$. Trying all eight possibilities for the first congruence and all nine for the second we discover that

$2x^2 + 5x - 3 \equiv 0 \pmod{8}$ has the unique solution $x \equiv 5 \pmod{8}$ and

$2x^2 + 5x - 3 \equiv 0 \pmod{9}$ has solutions $x \equiv 5, 6 \pmod{9}$.

The congruence $2x^2 + 5x - 3 \equiv 0 \pmod{72}$ therefore has two solutions:

(a) the simultaneous solution of $x \equiv 5 \pmod{8}$ and $x \equiv 5 \pmod{9}$, namely

$$x \equiv 5 \pmod{72}, \text{ and}$$

(b) the simultaneous solution of $x \equiv 5 \pmod{8}$ and $x \equiv 6 \pmod{9}$, namely

$$x \equiv 69 \pmod{72}. \qquad \blacklozenge$$

Problem 4.4 _____

Solve $x^2 - x + 4 \equiv 0 \pmod{80}$.

_____

We have now reduced the problem of solving polynomial congruences to the task of solving $P(x) \equiv 0 \left(\mathrm{mod}\ p^k \right)$ for prime p. In fact all solutions of this latter congruence can be built up from solutions of $P(x) \equiv 0 \pmod{p}$ in a way that we shall now illustrate.

Example 4.3

Solve $x^3 + x + 5 \equiv 0 \pmod{27}$.

As $27 = 3^3$, we first solve the given congruence modulo 3. Trying $x = 0, 1$ and 2 we find that

$$x^3 + x + 5 \equiv 0 \pmod{3}$$

has the unique solution $x \equiv 2 \pmod{3}$.

Moving up a power of 3, we next investigate

$$x^3 + x + 5 \equiv 0 \pmod{9}.$$

If a is a solution then $a^3 + a + 5$ is divisible by 9 (and therefore by 3) and so a is also a solution of $x^3 + x + 5 \equiv 0 \pmod{3}$. That is, $a \equiv 2 \pmod{3}$, and so the only possibilities are $a \equiv 2$, 5 or 8 $\pmod{9}$. Putting in these values:

$$2^3 + 2 + 5 = 15 \equiv 6 \pmod{9};$$
$$5^3 + 5 + 5 = 135 \equiv 0 \pmod{9};$$
$$(-1)^3 + (-1) + 5 = 3 \equiv 3 \pmod{9},$$

and so $x \equiv 5 \pmod{9}$ is the unique solution.

Moving up one more power of 3 and repeating the argument, any solution of

$$x^3 + x + 5 \equiv 0 \pmod{27}$$

must be a solution of $x^3 + x + 5 \equiv 0 \pmod{9}$ and hence must be congruent to 5 modulo 9. So we need only check the three candidates $x \equiv 5$, 14 and 23 $\pmod{27}$:

$$5^3 + 5 + 5 = 135 \equiv 0 \pmod{27};$$
$$14^3 + 14 + 5 = 2763 \equiv 9 \pmod{27};$$
$$(-4)^3 + (-4) + 5 = -63 \equiv 18 \pmod{27}.$$

We conclude that $x \equiv 5 \pmod{27}$ is the unique solution of the given congruence.   ♦

The strategy adopted in Example 4.3 will work for a general congruence of this form. Each solution of congruence $P(x) \equiv 0 \pmod{p^k}$ is of the form $b + rp^{k-1}$, where b is a solution of the congruence $P(x) \equiv 0 \pmod{p^{k-1}}$ and $0 \leq r < p$.

Problem 4.5 _____

Solve the congruences

(a) $x^2 + x - 7 \equiv 0 \pmod{25}$, and

(b) $x^2 + 4x - 1 \equiv 0 \pmod{125}$.

_____

5 FERMAT'S AND WILSON'S THEOREMS REVISITED

The results of Fermat's Little Theorem and Wilson's Theorems are illustrated in the solutions of two well-known combinatorial problems.

> **Question 1** *Having enough beads to permit unlimited use of each of n colours, how many different necklaces consisting of p beads can be made, where p is prime?*

Do we understand what is being asked? We shall assume that the necklaces are produced by first forming a string of p beads and then joining the ends. Our first step is therefore to count how many strings of p beads there are. Well that's fairly straightforward. There are n choices for the first bead, n choices for the second bead, and so on. In all there are n^p strings.

Notice that, of the n^p strings, n strings consist of beads of one colour alone (one such for each colour). Hence there are $n^p - n$ strings involving at least two colours. As the single colour necklaces are easily counted we shall put these to one side and concentrate on the ones involving two or more colours.

As one must expect, the $n^p - n$ strings counted will not result in $n^p - n$ different necklaces being produced. For example, the five strings of beads involving colours A, B and C illustrated in Figure 5.1 all result in the same necklace.

The strings in this example are obtained from one another by removing the bead on the left and replacing it on the right of the string. We shall call the operation of moving a bead from the left end to the right end a 'recycle'. Two strings which can be obtained from each other by a sequence of recycles clearly correspond to the same necklace.

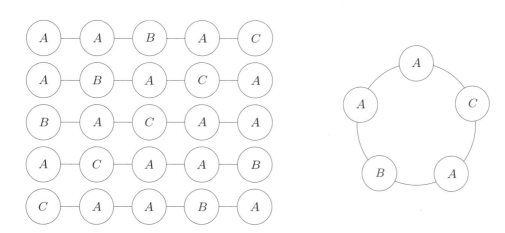

Figure 5.1   Five strings giving the same necklace

The five strings in Figure 5.1 are different. But if we were to consider the following string of four beads

Figure 5.2

we see that the second recycle would produce a string with the same colour scheme, BABA, and that further recycles here will produce just two different strings. So we need to ask how many different strings can be obtained from a string of p beads by recycles. The obvious answer is p, each of the p beads can occupy the left-end position, unless, as illustrated above for 4 beads, some fewer number of recycles restores the original colour scheme. We use the Division Algorithm to show that this cannot happen for a prime number of beads.

Let k be the least number of recycles which restores the colour scheme of a string of p beads, where $k > 1$ since we are ignoring strings of a single colour. Dividing p by k, there exist integers q and r such that

$$p = qk + r, \quad \text{where } 0 \leq r < k.$$

Now if k recycles restores the colour scheme then so too does $2k$, $3k$, $\ldots$, qk recycles. Furthermore p recycles certainly restores the colour scheme since each bead returns to its initial position. Therefore after the qkth recycle we have the original colour scheme and only a further r recycles are needed to restore it again. As k is the least number of recycles which can achieve this we must conclude that $r = 0$. This in turn implies that k divides p and, since p is prime and $k > 1$, we must have $k = p$.

The outcome of this is that the $n^p - n$ strings involving two or more colours are partitioned into disjoint sets of p strings, each set consisting of strings which can be obtained from each other by a sequence of recycles. This means that p divides $n^p - n$, which is Fermat's Little Theorem.

One might be tempted to conclude from this that there are $\dfrac{n^p - n}{p}$ different necklaces involving two or more colours and hence a grand total of $\dfrac{n^p - n}{p} + n$ different necklaces. Unfortunately there is a further complication to be taken into account. Consider the following string.

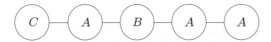

Figure 5.3

It is the first string of Figure 5.1 but with order reversed (i.e turned over left to right). It gives rise to a further set of five strings which correspond to the same necklace (turned over) as shown in Figure 5.1. So there are really ten different strings which produce this necklace. In general, two different sets of p strings can give rise to the same necklace. We say 'can' because turning a string over does not always produce a new string, as illustrated by this string:

Figure 5.4   A symmetrical string

Recycling gives five different strings but if we then turn it over and recycle we get the same five again. To complete the task we have still to count these 'symmetrical' strings. We leave you the task of completing Question 1 as a challenge. For the record, the number of necklaces is

$$\frac{n^p - n}{2p} + \frac{n^{(p+1)/2} - n}{2} + n.$$

Notice that as this is an integer value the first term still incorporates FLT.

A similar problem leads to a demonstration of Wilson's Theorem.

Question 2 *How many stellated p-gons are there, where p is prime?*

A stellated n-gon is formed by placing n points symmetrically around the circumference of a circle and then joining these points with n straight line segments, crossings being allowed. For $n = 5$ there are twelve stellated pentagons, as shown in Figure 5.5.

Notice that the five pentagons in the first row are congruent, being obtained from each other by rotations through $2\pi/5$. Similarly the pentagons in the second row are congruent. The remaining two are the *regular* stellated pentagons, having rotation through $2\pi/5$ as a symmetry operation. Hence, if we classify stellated pentagons up to similarity (for we are not concerned about the size) there are two essentially different irregular ones plus the two regular ones.

In the same way we now want to count the number of different stellated p-gons, where $p > 2$ is prime. We shall first count how many p-gons can be drawn (corresponding to the twelve pentagons) and then count how many of these are regular ones. The difference will give the irregular ones which will fall into sets of p congruent p-gons, being obtained from each other by rotation through $2\pi/p$.

To count the total number of stellated p-gons, imagine starting at any point on the circle and drawing the p-gon. There are $p - 1$ choices for the first destination, then $p-2$ choices for the second destination, and so on. In all $(p-1)!$ stellated p-gons can be drawn. But this has counted each p-gon twice because one ordering of the p vertices visited and the reverse order give the same p-gon. Hence there are $\dfrac{(p-1)!}{2}$ p-gons in all. (This agrees with the discovered 12 for the case $p=5$.)

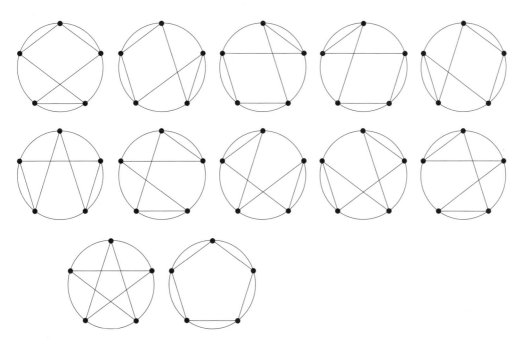

Figure 5.5   The twelve stellated pentagons

The regular stellated p-gons are formed by choosing any k in the range $1 \le k \le p-1$ and joining each point on the circle to the one k places round the circle clockwise from it. But again this counts each regular p-gon twice, because the choices of $k=r$ and $k=p-r$ will draw the same p-gon but visiting the points on the circle in reverse order. Hence there are $\dfrac{(p-1)}{2}$ regular stellated p-gons.

When p is not prime this construction will still draw a regular p-gon for each choice of k satisfying $\gcd(k,p)=1$.

Finally, the $\dfrac{(p-1)!}{2} - \dfrac{(p-1)}{2}$ irregular stellated p-gons are partitioned into congruent sets of p each as determined by rotation through $2\pi/p$. Hence the total number of different stellated p-gons is

$$\frac{(p-1)! - (p-1)}{2p} + \frac{p-1}{2}.$$

As this is an integer, the first term illustrates that p divides $(p-1)! - (p-1)$, or, in congruence notation, $(p-1)! \equiv -1 \pmod{p}$, which is Wilson's Theorem.

ADDITIONAL EXERCISES

Section 1

1   Find the least positive residue modulo 13 of 5^{12}, 5^{14} and 1996^{14}.

2 Use FLT to show that:

(a) $3^{50} + 5^{50}$ is divisible by 17;

(b) $2^{100} + 3^{100}$ is divisible by 97.

3 What is the remainder when 7^{154} is divided by 155? *Hint:* $155 = 5 \times 31$.

4 Show that $a^7 \equiv a \pmod{42}$ for all integers a.

5 If p and q are distinct primes, prove that $p^{q-1} + q^{p-1} \equiv 1 \pmod{pq}$.

6 Show that $(p-1)2^{p-1} + 1$ and $(p-2)2^{p-2}+1$ are each divisible by the odd prime p. Is it true that $(p-3)2^{p-3} + 1$ is necessarily divisible by the odd prime p?

Section 2

1 Leibnitz claimed, in 1677, that for any positive integer n with $\gcd(n, 10) = 1$, the length of the cycle of $\dfrac{1}{n}$ is a divisor of $n - 1$. Determine the decimal fraction for $\frac{1}{21}$ and prove him wrong.

2 From Theorem 2.2 (page 13) it can be deduced that for each positive integer n, there are at most a finite number of primes p for which the length of the cycle of $\dfrac{1}{p}$ is n. Explain how.

3 By first determining the prime decomposition of $10^4 - 1$, find all the primes p for which $\dfrac{1}{p}$ has a cycle of length 4. From the decomposition of $10^6 - 1$, find all primes p for which $\dfrac{1}{p}$ consists of a cycle of length 6.

4 Consider an ordinary pack of 52 playing cards. In the perfect faro shuffle the pack is cut into
two equal halves which are then placed (flicked) alternately, as illustrated.

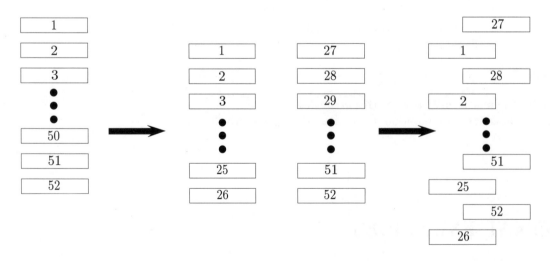

Figure 5.6   The perfect faro shuffle

The cards which initially occupy positions 1, 2, ..., 26 are moved to positions 2, 4, ..., 52 whilst the cards initially in positions 27, 28, ..., 52 are moved to positions 1, 3, ..., 51. Thus the card starting at position x moves to position y, where $1 \leq y \leq 52$ and

$y \equiv 2x \pmod{53}$. After n such shuffles this card will occupy position $2^n x \pmod{53}$.

(a) After how many perfect faro shuffles will the pack return to its initial order for the first time?

(b) If two jokers are included in the pack, making 54 cards in all, after how many perfect faro shuffles would this pack be restored to the original order for the first time?

Section 3

1 If a is a primitive root of p then the numbers a, a^2, $a^3, \ldots, a^{p-2}$ are congruent to the numbers 2, 3, 4, $\ldots$, $p-1$, in some order, and

$$a^k a^{p-k-1} = a^{p-1} \equiv 1 \pmod{p}.$$

Show that 2 is a primitive root of 29 and use the given observation to find the distribution of $\{2, 3, \ldots, 27\}$ into 13 pairs in such a way that the product of each pair is congruent modulo 29 to 1. Hence illustrate the proof of Wilson's Theorem for the prime 29.

2 If p is a prime and a is any integer, prove that p divides both

$$a^p + a \times (p-1)! \quad \text{and} \quad a + a^p \times (p-1)!$$

3 For p prime, prove that

$$(p-1)! \equiv p - 1 \pmod{1 + 2 + 3 + \cdots + (p-1)}.$$

Section 4

1 Show that $x = 1$ and $x = -2$ both satisfy

$$x^3 + 2x^2 + 6x + 5 \equiv 0 \pmod{7}.$$

Hence factorize $x^3 + 2x^2 + 6x + 5$ modulo 7 and find any further solutions of the congruence.

2 Solve the following polynomial congruences.
   (a) $x^7 + x^6 + x^5 + x^4 + x^3 + x^2 + x + 1 \equiv 0 \pmod{5}$
   (b) $x^{12} - 1 \equiv 0 \pmod{7}$
   (c) $x^5 - x^3 \equiv 0 \pmod{11}$
   (d) $x^5 + x^3 \equiv 0 \pmod{17}$

3 Solve the following polynomial congruences.
   (a) $x^3 + 2x^2 + 3x + 9 \equiv 0 \pmod{35}$
   (b) $x^3 + x^2 + 3x + 1 \equiv 0 \pmod{105}$
   (c) $x^2 - 7x + 31 \equiv 0 \pmod{75}$

Challenge Problems

1 Show that if n is not a multiple of 42 then $n^6 + 1091$ is composite. Show that the only possible values of n for which $n^6 + 1091$ is prime are $n = 210k$, or $n = 210k + 84$, or $n = 210k - 84$. In fact $n = 3906$ is the smallest value of n for which $n^6 + 1091$ is prime.

2   Prove that if the prime p satisfies $(p-1)! + 1 = p^r$, for some integer r, then $p = 2$, 3 or 5.

3   Prove that if $\gcd(n, 10) = 1$ then n^{101} terminates in the same three digits as does n.

4   Prove that if a and b are any integers and $p > 3$ is prime, then

$$ab^p - ba^p$$

is divisible by $6p$.

5   Prove that, if p is prime, then the binomial coefficient $^{2p}C_p$ is congruent modulo p to 2. (The binomial coefficients are defined by

$$^nC_r = \frac{n!}{r!(n-r)!}$$

for non-negative integers n and r with $r \leq n$.)

SOLUTIONS TO THE PROBLEMS

Solution 1.1

(a) Suppose that $a^p \equiv a \pmod{p}$ for all integers a. Then if $a \not\equiv 0 \pmod{p}$ we can cancel by a to obtain $a^{p-1} \equiv 1 \pmod{p}$, as required.

(b) Suppose that $a^{p-1} \equiv 1 \pmod{p}$ for all a such that $a \not\equiv 0 \pmod{p}$. Then, multiplying by a, $a^p \equiv a \pmod{p}$ for all $a \not\equiv 0 \pmod{p}$. However, if $a \equiv 0 \pmod{p}$ then $a^p \equiv a \equiv 0 \pmod{p}$ and so $a^p \equiv a \pmod{p}$ for all integers a.

Solution 1.2

(a) FLT tells us that $5^6 \equiv 1 \pmod{7}$. Therefore

$$5^{20} \equiv (5^6)^3 \times 5^2 \equiv 1^3 \times 25 \equiv 25 \equiv 4 \pmod{7},$$

and the remainder is 4.

(b) $37 \equiv 3 \pmod{17}$ and so

$$37^{37} \equiv 3^{37} \pmod{17}.$$

We first replace the 37 by its least positive residue before using FLT.

Now FLT tells us that $3^{16} \equiv 1 \pmod{17}$ and so

$$37^{37} \equiv 3^{37} \equiv (3^{16})^2 \times 3^5 \equiv 1^2 \times 3^5 \equiv 3^5 \pmod{17}$$
$$\equiv 27 \times 9 \equiv 10 \times 9 \equiv 90 \equiv 5 \pmod{17},$$

and the remainder is 5.

Solution 1.3

If $x \equiv a^{p-2}b \pmod{p}$ then $ax \equiv a^{p-1}b \pmod{p}$. Now, since $\gcd(a, p) = 1$, FLT gives $a^{p-1} \equiv 1 \pmod{p}$ and so $ax \equiv b \pmod{p}$, as claimed.

The congruence $5x \equiv 18 \pmod{19}$ has a unique solution modulo 19 which we now know is $x \equiv 5^{17} \times 18 \pmod{19}$. As $5^2 \equiv 6 \pmod{19}$ we have

$$x \equiv 5^{17} \times 18 \pmod{19}$$
$$\equiv (6^8 \times 5) \times (-1) \pmod{19}$$
$$\equiv (-2)^4 \times (-5) \pmod{19}, \quad \text{since } 6^2 \equiv -2 \pmod{19},$$
$$\equiv (-3) \times (-5) \pmod{19}$$
$$\equiv 15 \pmod{19}.$$

Solution 1.4

As $168 = 3 \times 7 \times 8$, to prove that 168 divides $a^6 - 1$, we shall show that $a^6 \equiv 1 \pmod 3$, $a^6 \equiv 1 \pmod 7$ and $a^6 \equiv 1 \pmod 8$. Since 3, 7 and 8 are relatively prime in pairs, it then follows from Theorem 1.3 of *Unit 3*, that $a^6 \equiv 1 \pmod{168}$.

Given that $\gcd(a, 42) = 1$ we know that $\gcd(a, 2) = 1$, $\gcd(a, 3) = 1$ and $\gcd(a, 7) = 1$.

From $\gcd(a, 7) = 1$, FLT gives $a^6 \equiv 1 \pmod 7$.

From $\gcd(a, 3) = 1$, FLT gives $a^2 \equiv 1 \pmod 3$, whereupon

$$a^6 \equiv (a^2)^3 \equiv 1^3 \equiv 1 \pmod 3.$$

Finally, from $\gcd(a, 2) = 1$ we have that a is odd. But

$$1^2 \equiv 3^2 \equiv 5^2 \equiv 7^2 \equiv 1 \pmod 8,$$

Notice that as 8 is not prime we cannot use FLT.

and so $a^2 \equiv 1 \pmod 8$.

Therefore $a^6 \equiv (a^2)^3 \equiv 1^3 \equiv 1 \pmod 8.$

Solution 1.5

To focus on the units digit of a number amounts to consideration of the number modulo 10. So our task is to show that $a^5 \equiv a \pmod{10}$.

Now FLT gives $a^5 \equiv a \pmod 5$. Moreover, $a^5 \equiv a \pmod 2$, since a^5 is even if, and only if, a is even. As 2 and 5 are relatively prime it follows that $a^5 \equiv a \pmod{10}$.

What can we now say about $a^{100} \pmod{10}$?

$$a^{100} \equiv (a^5)^{20} \equiv a^{20} \equiv (a^5)^4 \equiv a^4 \pmod{10},$$

and so a^{100} and a^4 have the same units digit.

Solution 2.1

(a)
$$1 = 0 \times 15 + 1$$
$$10 = 0 \times 15 + 10$$
$$100 = 6 \times 15 + 10$$
$$100 = 6 \times 15 + 10$$

This last equation now repeats indefinitely. So $\frac{1}{15} = 0.0\langle 6 \rangle$.

(b)
$$1 = 0 \times 41 + 1$$
$$10 = 0 \times 41 + 10$$
$$100 = 2 \times 41 + 18$$
$$180 = 4 \times 41 + 16$$
$$160 = 3 \times 41 + 37$$
$$370 = 9 \times 41 + 1$$
$$10 = 0 \times 41 + 10$$

The block of five equations now repeats indefinitely and so $\frac{1}{41}$ has the cycle length 5 and $\frac{1}{41} = 0.\langle 02439 \rangle$.

Solution 2.2

FLT tells us that $10^{16} \equiv 1 \pmod{17}$ and so the straightforward approach is to check each smaller positive power of 10 in turn looking for a least positive residue of 1. Going from one power to the next by multiplying by 10, the successive powers of 10 modulo 17 are:

$$10; \ 100 \equiv 15; \ 150 \equiv 14; \ 140 \equiv 4; \ 40 \equiv 6; \ 60 \equiv 9; \ 90 \equiv 5; \ 50 \equiv 16; \ 160 \equiv 7$$
$$70 \equiv 2; \ 20 \equiv 3; \ 30 \equiv 13; \ 130 \equiv 11; \ 110 \equiv 8; \ 80 \equiv 12; \ 120 \equiv 1.$$

The order of 10 modulo 17 is 16, and the cycle of $\frac{1}{17}$ has length 16.

Solution 2.3

We know that $10^{60} \equiv 1 \pmod{61}$ and that c is the smallest positive integer for which $10^c \equiv 1 \pmod{61}$. Suppose that dividing 60 by c gives $60 = cq + r$, where $0 \le r < c$.

Then $10^{60} = (10^c)^q \times 10^r \equiv 1^q \times 10^r \equiv 10^r \pmod{61}$. But $10^{60} \equiv 1 \pmod{61}$ so $10^r \equiv 1 \pmod{61}$. Therefore $r = 0$, for otherwise the order of 10 modulo 61 would be smaller than c. Hence c divides 60, as claimed.

Solution 2.4

Notice that 10 is a primitive root of 17 (from Table 2.2 or from the length of the cycle of $\frac{1}{17}$ being 16 and Theorem 2.2).

$$2 = 0 \times 17 + 2$$
$$20 = 1 \times 17 + 3$$
$$30 = 1 \times 17 + 13$$

At this point we know that $\frac{2}{17}$ begins 0.11 which pinpoints the starting point in the cycle of $\frac{1}{17}$. Thus

$$\tfrac{2}{17} = 0.\langle 1176470588235294 \rangle.$$

Similarly

$$9 = 0 \times 17 + 9$$
$$90 = 5 \times 17 + 5$$
$$50 = 2 \times 17 + 16$$

shows that $\frac{9}{17}$ begins 0.52, and so we can deduce that

$$\tfrac{9}{17} = 0.\langle 5294117647058823 \rangle.$$

Solution 3.1

(a) Wilson's Theorem assures us that 19 divides $18! + 1$. For any prime $p < 19$, p divides $18!$ and so p cannot divide $18! + 1$. Hence 19 is the smallest prime dividing $18! + 1$.

(b) Wilson's Theorem tells us that 31 divides $30! + 1$. Now

$$30! + 1 = 30 \times 29! + 1 \equiv (-1)29! + 1 \equiv 1 - 29! \equiv 0 \pmod{31},$$

and so 31 divides $29! - 1$. Furthermore 31 must be the smallest prime with this property because any prime $p < 31$ divides $29!$ and so cannot divide $29! - 1$.

Solution 3.2

Example 3.1 showed that for any prime $p = 4k + 1$ the congruence $x^2 + 1 \equiv 0 \pmod{4k + 1}$ has $x \equiv (2k)! \pmod{4k + 1}$ as a solution. For $k = 7$, the congruence $x^2 + 1 \equiv 0 \pmod{29}$ has solution $x \equiv 14! \pmod{29}$. Working modulo 29 we can simplify the calculation of $14!$ by pairing the terms appropriately:

This, of course, is not a very efficient method of solving the congruence. Its inclusion is to highlight the construction of Example 3.1

$$14! = 1 \times (2 \times 14) \times (3 \times 10) \times (4 \times 7) \times (5 \times 6) \times (8 \times 11) \times (9 \times 13) \times 12$$
$$\equiv 1 \times (-1) \times 1 \times (-1) \times 1 \times 1 \times 1 \times 12 \pmod{29}$$
$$\equiv 12 \pmod{29}.$$

Since $(-x)^2 = x^2$, $x \equiv -12 \equiv 17 \pmod{29}$ is a second solution.

In fact $x \equiv 12, 17 \pmod{29}$ are the only solutions of this congruence, as we shall see from work of the next section.

Solution 3.3

Let us first illustrate the proof for the particular case $p = 11$. Working modulo 11 we have

$$2 \equiv -9,\ 4 \equiv -7,\ 6 \equiv -5,\ 8 \equiv -3 \text{ and } 10 \equiv -1.$$

Therefore,

$$-1 \equiv 10! \pmod{11}$$
$$\equiv 1 \times 2 \times 3 \times 4 \times 5 \times 6 \times 7 \times 8 \times 9 \times 10 \pmod{11}$$
$$\equiv 1 \times 3 \times 5 \times 7 \times 9 \times (-9) \times (-7) \times (-5) \times (-3) \times (-1) \pmod{11}$$
$$\equiv (-1)^5 \times 1^2 \times 3^2 \times 5^2 \times 7^2 \times 9^2 \pmod{11}.$$

Multiplying through the final congruence by $(-1)^5$ gives

$$1^2 \times 3^2 \times 5^2 \times 7^2 \times 9^2 \equiv (-1)^6 \pmod{11},$$

which is one of the required formulae. The second comes quickly from the observation that from $2 \equiv -9 \pmod{11}$ we have $2^2 \equiv 9^2 \pmod{11}$ and similarly

$$4^2 \equiv 7^2,\ 6^2 \equiv 5^2,\ 8^2 \equiv 3^2 \text{ and } 10^2 \equiv 1^2 \pmod{11}.$$

Substitution gives

$$2^2 \times 4^2 \times 6^2 \times 8^2 \times 10^2 \equiv (-1)^6 \pmod{11}.$$

The same argument works for any odd prime p

$$2 \equiv -(p-2),\ 4 \equiv -(p-4),\ \ldots,\ (p-1) \equiv -1 \pmod{p},$$

and so

$$-1 \equiv (p-1)! \pmod{p}$$
$$= (1 \times 3 \times 5 \times \cdots \times (p-2)) \times (2 \times 4 \times 6 \times \cdots \times (p-1)) \pmod{p}$$
$$\equiv (1 \times 3 \times 5 \times \cdots \times (p-2)) \times (-(p-2) \times (-(p-4)) \times \cdots \times (-1)) \pmod{p}$$
$$\equiv (-1)^{(p-1)/2} \times 1^2 \times 3^2 \times \cdots \times (p-2)^2 \pmod{p}.$$

Multiplying by $(-1)^{(p-1)/2}$ gives

$$1^2 \times 3^2 \times \cdots \times (p-2)^2 \equiv (-1)^{(p+1)/2} \pmod{p}.$$

For each k on the left of this congruence we can replace k^2 by $(p-k)^2$ to get the alternative formula

$$2^2 \times 4^2 \times \cdots \times (p-1)^2 \equiv (-1)^{(p+1)/2} \pmod{p}.$$

Solution 4.1

(a) Trying $x = 0, 1, 2, 3, 4$ and 5, we find that $x^3 \equiv x \pmod{6}$ in each case and so the congruence $x^3 - x \equiv 0 \pmod{6}$ has six solutions, namely $x \equiv 0, 1, 2, 3, 4, 5 \pmod{6}$.

(b) The straightforward approach is to test the result as x runs through a complete set of residues. However, recalling that $x^2 - 1 \equiv 0 \pmod{8}$ See Solution 1.4. whenever x is odd, it follows that

$$x^3 - x = x(x^2 - 1) \equiv 0 \pmod{8} \text{ for } x \equiv 1, 3, 5, 7 \pmod{8}.$$

For x even, $0^3 - 0 \equiv 0$, $2^3 - 2 \equiv 6$, $4^3 - 4 \equiv 4$ and $6^3 - 6 \equiv 2 \pmod{8}$.

Hence there are five solutions in all, $x \equiv 0, 1, 3, 5, 7 \pmod{8}$.

(c) The values of the polynomial for a complete set of residues are given below.

x	$x^2 + x + 1$ (mod 7)
0	1
1	3
2	$7 \equiv 0$
3	$13 \equiv 6$
4	$21 \equiv 0$
5	$31 \equiv 3$
6	$43 \equiv 1$

The numbers in the calculations here could have been kept smaller by using the set of least absolute residues modulo 7.

There are two solutions, namely $x \equiv 2, 4$ (mod 7).

(d) First, we simplify the congruence by reducing the coefficients modulo 7 to get the equivalent congruence $2x^2 + 2x + 1 \equiv 0$ (mod 7). As x takes values 0, 1, 2, 3, 4, 5 and 6, $2x^2 + 2x + 1$ takes respective values (modulo 7) 1, 5, 6, 4, 6, 5 and 1. Therefore this congruence has no solutions.

Solution 4.2

(a) This has degree 6. You may have noticed that this congruence can be reduced by FLT to an equivalent congruence of lower degree (0 in fact), but this is not immediately pertinent.

(b) Since $10 \equiv 0$ (mod 5), the leading term modulo 5 is $3x$, so the congruence has degree 1.

(c) Since $15 \equiv 0$ (mod 5) and $-10 \equiv 0$ (mod 5), the leading term modulo 5 is the constant 2, so the congruence has degree 0.

Solution 4.3

(a) Evaluating $x^2 + 4x - 4$ for $x = 0, \pm 1, \pm 2$ and ± 3, two solutions $x \equiv -1$ and $x \equiv -3$ are found. Pulling out the factor $(x + 1)$ corresponding to the first of these solutions,

$$x^2 + 4x - 4 \equiv (x + 1)(Ax + B) \text{ (mod 7)}$$

produces $A = 1$, $B = 3$ and the factorization

$$x^2 + 4x - 4 \equiv (x + 1)(x + 3) \text{ (mod 7)}.$$

The factors $x + 1$ and $x + 3$ could equally well be written as $x - 6$ and $x - 4$ respectively.

The second factor is the one expected, corresponding to solution $x \equiv -3$ (mod 7).

(b) By exhaustion, we find that $x^3 + 2x^2 + x + 3 \equiv 0$ (mod 7) has just the two solutions $x \equiv 1, 2$ (mod 7).

Pulling out the factor $(x - 1)$:

$$x^3 + 2x^2 + x + 3 \equiv (x - 1)(x^2 + 3x + 4) \text{ (mod 7)},$$

and then pulling the factor $(x - 2)$ out gives

$$x^3 + 2x^2 + x + 3 \equiv (x - 1)(x - 2)(x + 5) \text{ (mod 7)}$$
$$\equiv (x - 1)(x - 2)^2 \text{ (mod 7)}.$$

Solution 4.4

As $80 = 2^4 \times 5$ we solve the congruences $x^2 - x + 4 \equiv 0$ (mod 16) and $x^2 - x + 4 \equiv 0$ (mod 5). By exhaustion

$$x^2 - x + 4 \equiv 0 \text{ (mod 16)}$$

has solutions $x \equiv -3, 4$ (mod 16), and

$$x^2 - x + 4 \equiv 0 \text{ (mod 5)}$$

has the unique solution $x \equiv 3$ (mod 5).

Hence $x^2 - x + 4 \equiv 0 \pmod{80}$ has two solutions. By solving simultaneously,

$$x \equiv -3 \pmod{16} \text{ and } x \equiv 3 \pmod 5 \text{ yield } x \equiv 13 \pmod{80}$$

and

$$x \equiv 4 \pmod{16} \text{ and } x \equiv 3 \pmod 5 \text{ yield } x \equiv 68 \pmod{80}.$$

Solution 4.5

(a) By trial and error we find that $x^2 + x - 7 \equiv 0 \pmod 5$ has solutions $x \equiv 1, 3 \pmod 5$. So the only candidates for solution of $x^2 + x - 7 \equiv 0 \pmod{25}$ are:

> from $x \equiv 1 \pmod 5$, $\quad x \equiv 1, 6, 11, 16 \text{ and } 21 \pmod{25}$

and

> from $x \equiv 3 \pmod 5$, $\quad x \equiv 3, 8, 13, 18 \text{ and } 23 \pmod{25}$.

Of these

$$11^2 + 11 - 7 = 125 \equiv 0 \pmod{25}$$

and

$$13^2 + 13 - 7 = 175 \equiv 0 \pmod{25}$$

are the only solutions. Hence $x^2 + x - 7 \equiv 0 \pmod{25}$ has solutions $x \equiv 11, 13 \pmod{25}$.

As 5 is prime we know by Lagrange's Theorem there are at the most two solutions; once we have found them we can stop.

This method still involves much computation, but whereas exhaustion would involve testing 25 values modulo 25, we have reduced this number to 10.

(b) The congruence $x^2 + 4x - 1 \equiv 0 \pmod 5$ has unique solution $x \equiv 3 \pmod 5$. The candidates for solution of $x^2 + 4x - 1 \equiv 0 \pmod{25}$ are therefore $x \equiv 3, 8, 13, 18 \text{ and } 23 \pmod{25}$.

But

$$3^2 + 12 - 1 = 20 \equiv 20 \pmod{25};$$
$$8^2 + 32 - 1 = 95 \equiv 20 \pmod{25};$$
$$13^2 + 52 - 1 = 220 \equiv 20 \pmod{25};$$
$$(-7)^2 - 28 - 1 \equiv 20 \pmod{25};$$
$$(-2)^2 - 8 - 1 = -5 \equiv 20 \pmod{25}.$$

Hence $x^2 + 4x - 1 \equiv 0 \pmod{25}$ has no solutions, and from this $x^2 + 4x - 1 \equiv 0 \pmod{125}$ can have no solutions.

This time we have saved a lot of computation!

SOLUTIONS TO ADDITIONAL EXERCISES

Section 1

1 $5^{12} \equiv 1 \pmod{13}$ by FLT.

$5^{14} = 5^{12} \times 5^2 \equiv 1 \times 5^2 \equiv 25 \equiv 12 \pmod{13}$.

$1996^{14} \equiv 7^{14} \pmod{13}$,   since $1996 \equiv 7 \pmod{13}$,
$\equiv 7^{12} \times 7^2 \equiv 1 \times 7^2 \pmod{13}$,   since $7^{12} \equiv 1 \pmod{13}$ by FLT,
$\equiv 10 \pmod{13}$.

2 (a) FLT gives $3^{16} \equiv 1 \pmod{17}$ and $5^{16} \equiv 1 \pmod{17}$. Therefore

$$3^{50} = (3^{16})^3 3^2 \equiv 1^3 \times 9 \equiv 9 \pmod{17}$$

and

$$5^{50} = (5^{16})^3 5^2 \equiv 1^3 \times 25 \equiv 8 \pmod{17}.$$

Hence $3^{50} + 5^{50} \equiv 9 + 8 \equiv 0 \pmod{17}$, and hence $3^{50} + 5^{50}$ is divisible by 17.

(b) As 97 is prime, FLT gives $2^{96} \equiv 3^{96} \equiv 1 \pmod{97}$. Therefore

$$2^{100} + 3^{100} = 2^{96} \times 2^4 + 3^{96} \times 3^4 \equiv 2^4 + 3^4 \equiv 16 + 81 \equiv 0 \pmod{97}.$$

Hence $2^{100} + 3^{100}$ is divisible by 97.

3 $155 = 5 \times 31$, and FLT gives $7^4 \equiv 1 \pmod 5$ and $7^{30} \equiv 1 \pmod{31}$. Therefore

$$7^{154} = (7^{30})^5 \times 7^4 \equiv 7^4 \equiv 18^2 \equiv 14 \pmod{31}$$

and

$$7^{154} = (7^4)^{38} \times 7^2 \equiv 7^2 \equiv 4 \pmod 5.$$

By the Chinese Remainder Theorem, the simultaneous congruences $x \equiv 14 \pmod{31}$ and $x \equiv 4 \pmod 5$ have a unique solution modulo 155 which can be seen to be $x \equiv 14 \pmod{155}$. So the remainder on dividing 7^{154} by 155 is 14.

4 As $42 = 2 \times 3 \times 7$, we shall show that $a^7 \equiv a \pmod 2$, $a^7 \equiv a \pmod 3$ and $a^7 \equiv a \pmod 7$, and then appeal to the corollary to Theorem 1.3 of *Unit 3*.

Let a be *any* integer. For each of the primes $p = 2$, 3 and 7 we apply FLT in the form $a^p \equiv a \pmod p$.

$$p = 2 : \quad a^7 = (a^2)^3 a \equiv a^3 a \equiv (a^2)^2 \equiv a^2 \equiv a \pmod 2$$
$$p = 3 : \quad a^7 = (a^3)^2 a \equiv a^2 a \equiv a^3 \equiv a \pmod 3$$
$$p = 7 : \quad a^7 \equiv a \pmod 7$$

Hence, $a^7 \equiv a \pmod{42}$ for all integers a.

5 By FLT, $p^{q-1} \equiv 1 \pmod q$ and, since q divides q^{p-1},

$$p^{q-1} + q^{p-1} \equiv 1 \pmod q.$$

Similarly

$$p^{q-1} + q^{p-1} \equiv 1 \pmod p.$$

Therefore, since $\gcd(p, q) = 1$,

$$p^{q-1} + q^{p-1} \equiv 1 \pmod{pq}.$$

6 For odd prime p we have $2^{p-1} \equiv 1 \pmod p$. Therefore

$$(p - 1)2^{p-1} + 1 \equiv (p - 1) \times 1 + 1 \equiv p \equiv 0 \pmod p$$

and

$$(p - 2)2^{p-2} + 1 = p \times 2^{p-2} - 2^{p-1} + 1 \equiv 0 - 1 + 1 \equiv 0 \pmod p.$$

There is no obvious reason why $(p - 3)2^{p-3} + 1$ should be divisible by p, and indeed $p = 3$ gives a counter-example:

$$(3 - 3)2^{3-3} + 1 = 1 \not\equiv 0 \pmod 3.$$

Section 2

1 The system of equations for the decimal fraction of $\frac{1}{21}$, is

$$1 = 0 \times 21 + 1$$
$$10 = 0 \times 21 + 10$$
$$100 = 4 \times 21 + 16$$
$$160 = 7 \times 21 + 13$$
$$130 = 6 \times 21 + 4$$
$$40 = 1 \times 21 + 19$$
$$190 = 9 \times 21 + 1$$

The equations now cycle. So $\frac{1}{21} = 0.\langle 047619 \rangle$, having a cycle of length 6. As 6 not a divisor of $21 - 1$, Leibnitz' claim is disproved.

2 If $\dfrac{1}{p}$ has a cycle of length n then $10^n \equiv 1 \pmod{p}$. That is, p divides $10^n - 1$. Hence, $p < 10^n - 1$, confirming that the number of such p is finite.

3 The prime decomposition of $10^4 - 1$ is

$$10^4 - 1 = 3^2 \times 11 \times 101.$$

If $\dfrac{1}{p}$ has a cycle of length 4 then $10^4 \equiv 1 \pmod{p}$ and so p must be a divisor of $10^4 - 1$. These values are seen to be $p = 3$, 11, and 101. For each of these p the order of 10 modulo p is a divisor of 4, namely 1, 2 or 4. Those p for which the order is less than 4 must satisfy $10^2 \equiv 1 \pmod{p}$, which happens only for p a divisor of $10^2 - 1 = 99$; that is, $p = 3$ or 11. We conclude that the only prime p for which $\dfrac{1}{p}$ has a cycle of length 4 is 101.

In the same way, those p for which $\dfrac{1}{p}$ has a cycle of length 6 are those prime divisors of $10^6 - 1$ which are not divisors of either $10^2 - 1$ or $10^3 - 1$. That is, any divisor of $3^3 \times 7 \times 11 \times 13 \times 37$ which is not a divisor of either $3^2 \times 11$ or $3^3 \times 37$. The only primes satisfying this condition are 7 and 13.

4 (a) If all cards return to their original position after n shuffles then

$$2^n x \equiv x \pmod{53}, \text{ for all } 1 \le x \le 52.$$

As $\gcd(x, 53) = 1$ we may cancel x to obtain

$$2^n \equiv 1 \pmod{53}.$$

So n has to be a multiple of the order of 2 modulo 53, and indeed if n shuffles is the smallest number for which this occurs, n is the order of 2 modulo 53.

FLT tells us that $2^{52} \equiv 1 \pmod{53}$ so, by Theorem 2.3, the n we seek is a divisor of 52. Checking $n = 2$, 4, 13 and 26 we find

$$2^2 = 4;$$
$$2^4 = 16;$$
$$2^{13} = 2^6 \times 2^6 \times 2 \equiv 11 \times 11 \times 2 \equiv 30 \pmod{53};$$
$$2^{26} = (2^{13})^2 \equiv 30^2 \equiv 52 \pmod{53}.$$

Therefore, as no smaller power of 2 is congruent modulo 53 to 1, the order of 2 modulo 53 is 52, and 52 shuffles are needed to restore the original order.

(b) When the number of cards in the pack goes up to 54 we have to find the smallest positive integer n for which $2^n \equiv 1 \pmod{55}$. We note that 55 is not prime so we are not able to start from $2^{54} \equiv 1 \pmod{55}$. However noting that $55 = 5 \times 11$, FLT gives

$$2^4 \equiv 1 \pmod 5 \text{ and } 2^{10} \equiv 1 \pmod{11}.$$

As $\mathrm{lcm}(4, 10) = 20$ we have

$$2^{20} = (2^4)^5 \equiv 1 \pmod 5 \text{ and } 2^{20} = (2^{10})^2 \equiv 1 \pmod{11}.$$

Since 5 and 11 are relatively prime the corollary to Theorem 1.3 of *Unit 3* gives $2^{20} \equiv 1 \pmod{55}$, and we have one solution to our problem; 20 shuffles will restore the original order of the pack of 54 cards.

If c is the smallest positive integer such that $2^c \equiv 1 \pmod{55}$ then c must divide 20. Why? Suppose $20 = qc + r$, where $0 \le r < c$. Then, as in the proof of Theorem 2.3,

$$a^{20} \equiv a^{qc+r} \equiv (a^c)^q \times a^r \equiv 1^q a^r \pmod{55}$$

showing that $a^r \equiv 1 \pmod{55}$ with $r < c$. Hence $r = 0$ and so c must divide 20. However,

$$2^2 = 4;$$
$$2^4 = 16;$$
$$2^5 = 32;$$
$$2^{10} = 2^6 \times 2^4 \equiv 9 \times 16 \equiv 144 \equiv 34 \pmod{55},$$

shows that no smaller power of 2 is congruent modulo 55 to 1. So 20 is the least number of shuffles which restores the original order.

The fact that 55 is not a prime does not affect the necessity for n to satisfy $2^n \equiv 1 \pmod{55}$ as $2^n x \equiv x \pmod{55}$ has to hold for *all* x including $x = 1$.

Section 3

1   The successive powers of 2 modulo 29 are given in the table:

n	2^n	n	2^n	n	2^n	n	2^n
1	2	8	24	15	27	22	5
2	4	9	19	16	25	23	10
3	8	10	9	17	21	24	20
4	16	11	18	18	13	25	11
5	3	12	7	19	26	26	22
6	6	13	14	20	23	27	15
7	12	14	28	21	17	28	1

To get from one power to the next we just 'double modulo 29'.

Note that 2^n takes all 28 non-zero values modulo 29, confirming that 2 is a primitive root of 29. Now, if $m + n = 28$ then $2^m 2^n = 2^{m+n} = 2^{28} \equiv 1 \pmod{29}$. For example, taking $m = 8$ and $n = 20$,

$$2^8 \times 2^{20} \equiv 24 \times 23 \equiv (-5) \times (-6) \equiv 30 \equiv 1 \pmod{29}.$$

In all there are 13 such pairs, 1 with 27, 2 with 26, ... and 13 with 15.

The value $n = 14$, (for which $2^{14} \equiv 28 \equiv -1 \pmod{29}$), is omitted because it pairs with itself. The pairs are

$$2 \times 15 \equiv 4 \times 22 \equiv 8 \times 11 \equiv 16 \times 20 \equiv 3 \times 10 \equiv 6 \times 5 \equiv 12 \times 17$$
$$24 \times 23 \equiv 19 \times 26 \equiv 9 \times 13 \equiv 18 \times 21 \equiv 7 \times 25 \equiv 14 \times 27 \equiv 1 \pmod{29}.$$

Multiplying all these thirteen pairs of numbers together gives

$$27! \equiv 1 \pmod{29},$$

and multiplying through this by 28 we reach

$$28! \equiv 28 \equiv -1 \pmod{29},$$

completing the illustration of the proof.

2   For any integer a and prime p, FLT gives $a^p \equiv a \pmod{p}$ and Wilson's Theorem gives $(p-1)! \equiv -1 \pmod{p}$. Therefore

$$a^p + a \times (p-1)! \equiv a + a \times (-1) \equiv 0 \pmod{p}$$

and

$$a + a^p \times (p-1)! \equiv a + a \times (-1) \equiv 0 \pmod{p}.$$

3   The result is certainly true for $p = 2$, so we assume that p is odd. Since

$$1 + 2 + 3 + \cdots + (p-1) = \frac{p(p-1)}{2},$$

and p and $\dfrac{p-1}{2}$ are relatively prime integers, it suffices to show that

$$(p-1)! \equiv p-1 \pmod{p}$$

and

$$(p-1)! \equiv p-1 \left(\bmod \ \frac{p-1}{2}\right).$$

The first of these is precisely Wilson's Theorem and the second simply says that $0 \equiv 0$ since $\dfrac{p-1}{2}$ is a divisor of $p-1$, and so

$$(p-1)! \equiv p-1 \left(\bmod \ \frac{p(p-1)}{2}\right).$$

Section 4

1   When $x = 1$

$$x^3 + 2x^2 + 6x + 5 = 1 + 2 + 6 + 5 = 14 \equiv 0 \pmod{7}$$

and when $x = -2$

$$x^3 + 2x^2 + 6x + 5 = -8 + 8 - 12 + 5 = -7 \equiv 0 \pmod{7}.$$

So both are solutions and it follows that

$$x^3 + 2x^2 + 6x + 5 \equiv (x-1)(x+2)(Ax+B) \pmod{7}.$$

A little algebra reveals that $A \equiv B \equiv 1 \pmod{7}$ and consequently $x + 1$ is also a factor. That gives a third solution, $x \equiv -1 \pmod{7}$. As Lagrange's Theorem guarantees that this congruence of degree 3 cannot have more than three solutions we conclude that $x \equiv 1, -2, -1 \pmod{7}$ are all the solutions.

The least positive residue solutions are $x \equiv 1,\ 5,\ 6 \pmod{7}$.

2   (a) Since $x \equiv 0 \pmod{5}$ is not a solution we may assume that $\gcd(5, x) = 1$ so that $x^4 \equiv 1 \pmod{5}$. Substituting this into the polynomial congruence produces

$$2(x^3 + x^2 + x + 1) \equiv 0 \pmod{5}.$$

Trial and error reveals that $x \equiv 2, 3$ and $4 \pmod{5}$ are the only three solutions.

(b) $x^{12} - 1 = (x^6 - 1)(x^6 + 1) \equiv 0 \pmod 7$.

$x \equiv 0 \pmod 7$ is not a solution, but for all other x we know from FLT that $x^6 - 1 \equiv 0 \pmod 7$. So the congruence has six solutions, namely $x \equiv 1, 2, 3, 4, 5$ and $6 \pmod 7$.

(c) $x^5 - x^3 = x^3(x^2 - 1) \equiv 0 \pmod{11}$.

$x^3 \equiv 0 \pmod{11}$ has the unique solution $x \equiv 0 \pmod{11}$, whilst $x^2 - 1 \equiv 0 \pmod{11}$ has at most two solutions (by Lagrange's Theorem) which can be seen to be $x \equiv \pm 1 \pmod{11}$. So there are three solutions, $x \equiv 0, 1$ and $10 \pmod{11}$.

If $x^3 \equiv 0 \pmod{11}$ then 11 divides x^3. But that can only happen when 11 divides x.

(d) $x^5 + x^3 = x^3(x^2 + 1) \equiv 0 \pmod{17}$.

$x^3 \equiv 0 \pmod{17}$ has the unique solution $x \equiv 0 \pmod{17}$, whilst $x^2 + 1 \equiv 0 \pmod{17}$ has at most two solutions (by Lagrange's Theorem). Writing the latter in the equivalent form $x^2 \equiv 16 \pmod{17}$ the solutions are seen to be $x \equiv \pm 4 \pmod{17}$. So there are three solutions, $x \equiv 0, 4$ and $13 \pmod{17}$.

3 (a) $x^3 + 2x^2 + 3x + 9 \equiv 0 \pmod{5 \times 7}$.

We solve the congruence modulo 5 and modulo 7 and then look for simultaneous solutions.

$$x^3 + 2x^2 + 3x + 4 \equiv 0 \pmod 5$$

has solution (found by exhaustion) $x \equiv 1 \pmod 5$.

$$x^3 + 2x^2 + 3x + 2 \equiv 0 \pmod 7$$

has solutions $x \equiv 3, 6 \pmod 7$.

$x \equiv 1 \pmod 5$ and $x \equiv 3 \pmod 7$ if, and only if, $x \equiv 31 \pmod{35}$.

$x \equiv 1 \pmod 5$ and $x \equiv 6 \pmod 7$ if, and only if, $x \equiv 6 \pmod{35}$.

There are two solutions, $x \equiv 6, 31 \pmod{35}$.

(b) $x^3 + x^2 + 3x + 1 \equiv 0 \pmod{3 \times 5 \times 7}$.

Proceeding as in part (a), we first solve the congruence for each of the moduli 3, 5 and 7. However, this congruence has no solutions modulo 5 (as discovered by trying each of the values $0, \pm 1, \pm 2$ for x). Hence the congruence can have no solutions modulo 105.

(c) As $75 = 3 \times 25$, we first solve $x^2 - 7x + 31 \equiv 0 \pmod 3$, then solve $x^2 - 7x + 31 \equiv 0 \pmod{25}$ and finally find simultaneous solutions.

Working modulo 3:

$$x^2 - 7x + 31 \equiv x^2 - x + 1 \equiv 0 \pmod 3 ,$$

which is seen to have the unique solutions $x \equiv 2 \pmod 3$.

To solve $x^2 - 7x + 31 \equiv 0 \pmod{25}$ we first find the solutions, if any, modulo 5. As

$$x^2 - 7x + 31 \equiv x^2 - 2x + 1 \equiv (x - 1)^2 \equiv 0 \pmod 5 ,$$

this has the unique solution $x \equiv 1 \pmod 5$. It follows that the only candidates for solutions modulo 25 are

$$x \equiv 1, 6, 11, 16, 21 \pmod{25} .$$

A little arithmetic confirms that each of these five is a solution.

Hence the solutions of the original congruence are found by solving $x \equiv 2 \pmod 3$ simultaneously with each of these five values modulo 25. Five solutions emerge.

$$x \equiv 11, 26, 41, 56, 71 \pmod{75} .$$

INDEX